A First Course in Mechanics

A First Course in Mechanics

Mary Lunn

St Hugh's College, Oxford

OXFORD · NEW YORK · TOKYO
OXFORD UNIVERSITY PRESS

Oxford University Press, Walton Street, Oxford OX2 6DP

Oxford New York
Athens Auckland Bangkok Bombay
Calcutta Cape Town Dar es Salaam Delhi
Florence Hong Kong Istanbul Karachi
Kuala Lumpur Madras Madrid Melbourne
Mexico City Nairobi Paris Singapore
Taipei Tokyo Toronto

and associated companies in
Berlin Ibadan

Oxford is a trade mark of Oxford University Press

Published in the United States
by Oxford University Press Inc., New York

First published 1991
Reprinted 1992, 1993, 1996

British Library Cataloguing in Publication Data
Lunn, M.
A first course in mechanics.
1. Mechanics
I. Title
531
ISBN 0 19 853430 2 (Hbk)
ISBN 0 19 853433 7 (Pbk)

Library of Congress Cataloging in Publication Data
Lunn, Mary
A first course in mechanics/Mary Lunn.
1. Mechanics, Analytic, I. Title
QA807.L95   1991   531–dc20   90–43228

Printed in Great Britain by
Bookcraft (Bath) Ltd.

To Dan, Rebecca, and Peter

Preface

Over the years the approach to teaching mechanics in university courses has been transformed by the use of modern algebra. Compact techniques are now accessible to students with a good basic understanding of vector and matrix methods. The aim of this book is to enable first- and second-year university students to use modern mathematical tools to study the rich model built on Newton's exposition of the physical world.

The book begins with a short look at one-dimensional systems so that a student with some mathematical expertise but no previous knowledge of mechanics as such may start here. Some knowledge of geometric vectors is assumed. There are many excellent books covering mathematical methods including use of vectors. The first part of the text is devoted to Newtonian vector mechanics, culminating in the theory of the motion of rigid bodies; the many examples include the gyrocompass. In later chapters the theory of Lagrangian mechanics is developed and in particular used to establish the behaviour of simple mechanical systems undergoing small oscillations about a position of stable equilibrium. The final chapter brings together all aspects of the theory to provide a concise treatment of impulsive motion. The examples are taken from astronomy, sport, the motion of artificial satellites, tops.

In two topics, namely conservation of energy and the inertia of a rigid body, a complete understanding of theory involves familiarity with vector analysis, symmetric matrices, or both. I have tried to organize the material so that those students who have not covered these topics can understand the model in sufficient depth to apply it without fear to practical systems. The most sophisticated use of linear algebra arises in the treatment of oscillations, a topic which would normally belong in a second-year undergraduate course.

Inspiration for the book came from two sources, these being first- and second-year courses in Oxford and a second-level applied mathematics course from the Open University. Teaching students from both universities has amplified and refined my own understanding and appreciation of Newton's mechanical model of the physical world. I would like to thank my colleagues and students, both at Oxford University and the Open University – colleagues for discussion and encouragement, students for unwittingly struggling through many examples.

The exercises concluding each chapter contain some questions taken from examination papers set in Oxford. The author wishes to thank Oxford University Press for permission to use them.

Oxford   M. L.
May 1990

Contents

1 Newton's laws

1.1 Introduction

Prior to the late nineteenth and early twentieth centuries the behaviour of the physical world was thought to be completely explained by the model proposed by Newton. In the sixteenth century Tycho Brahe made detailed and accurate astronomical observations which Johannes Kepler was able to analyse. Kepler's laws are in fact a mathematical description of the motion of the planets. They are:

1. The path of each planet is an ellipse with the Sun at a focus.

2. A straight line joining the Sun and a planet sweeps out equal areas in equal times.

3. The square of each planet's period is proportional to the cube of the semi-major axis of its elliptic orbit.

These laws explain geometrically how the planets move in relation to the Sun, but they do not enable us to make predictions about the motion of other astronomical bodies, for example a binary star system.

Newton developed a model based on laws of the physical world which was applied successfully both to mechanical systems situated on the Earth and to the motion of astrophysical bodies. Using a new mathematical tool called calculus he could derive Kepler's laws as a consequence. He could make predictions about the behaviour of other mechanical systems. Most of Newton's work on these laws was done during the years of the plague in 1665 and 1666 when he left Cambridge for the countryside to escape infection. It was published many years later in *Principia mathematica philosophiae naturalia* (1686). Newton's Laws are:

1. Every particle persists in its state of rest or of uniform motion in a straight line unless it is compelled to change that state by impressed forces.

2. The rate of change of motion is proportional to the motive force impressed; and is made in the direction of the straight line in which that force is impressed.

3. To every action there is an equal reaction: or the mutual actions of two bodies upon each other are always equal but oppositely directed.

Newton also concluded that 'there is a power of gravity pertaining to all bodies proportional to the several quantities of matter which they contain,' and again this 'propagates its virtue on all sides to immense distance, decreasing always as the inverse square of the distances'.

Until the early part of this century this model of the physical world was thought to apply to all mechanical systems. We now know that it does not provide a good model for very small objects, say atoms with a diameter of 10^{-10} m, where quantum mechanics comes into its own. Nor does it provide a satisfactory model for very fast moving objects, such as electrons moving in a television tube with a speed of 10^8 m s^{-1}, or for the huge masses and distances of the astrophysical world where Einstein's theory of relativity comes into play. Nevertheless, Newton's model of physical systems remains an important part of modern science, particularly since man acquired the ability to launch rockets and put satellites into space. The laws which govern the behaviour of the planets also govern the behaviour of artificial satellites.

1.2 Space, time, and paths

We need a geometrical framework in which to formulate Newton's laws. We assume that:

1. Space can be described by a system of three-dimensional geometric position vectors measured from some conveniently chosen origin. From this origin select a set of perpendicular Cartesian axes. This will be our *frame of reference*.

2. Time can be measured using a clock. If we set up an experiment and repeat it under identical conditions (e.g. run all the water out of a tank) we expect the time measured on a clock to be the same. In ancient times water clocks and sandglasses were used to measure time intervals.

We shall use the system of SI units which measures length in metres and time in seconds. The *mass* of an object can be found by comparison with other objects of known mass, for example by using a balance if the object is small enough. Such measurements are made every day in cooking using kitchen scales! In SI units the unit of mass is 1 kg (kilogram). Mass is a scalar quantity having magnitude but not direction.

A particle can move its position in space. How do we describe that motion and, even more fundamentally, what do we mean by a particle?

Definition 1.1
A particle is a mathematical object with a mass but without size or shape. Physically no such object exists but none the less in many systems a heavy object may have a size which is negligible compared with other lengths in the system. For example, we can consider the Earth as a particle for the purposes of obtaining a good approximation to its path around the Sun but the approximation is less good when investigating the relative motion of the Moon and Earth.

Let us suppose that we have chosen a suitable origin from which to locate the position of a particle. The particle has position vector **r** measured from that origin. As the particle changes its *position* with time we have

$$\mathbf{r} = \mathbf{r}(t)$$

i.e. in general the path is a three-dimensional curve, position being dependent on time.

In this geometrical framework we can describe Newton's concept of alteration of motion or *acceleration*.

Definition 1.2
The *velocity* of the particle is then the rate of change of position with time:

$$\mathbf{v} = \frac{d\mathbf{r}}{dt} (= \dot{\mathbf{r}})$$

and the *acceleration* is the rate of change of velocity with time:

$$\mathbf{a} = \frac{d^2\mathbf{r}}{dt^2} = \ddot{\mathbf{r}}.$$

Velocity and acceleration are vector quantities having both magnitude and direction. A dot over a quantity will always mean differentiation with respect to time.

Consider the following simple examples.

Example 1
Suppose a particle moves in straight line with constant acceleration a. Then if we measure **r** from the particle's position at time $t = 0$ and the initial velocity is known to be $U\mathbf{i}$, where **i** is a constant unit vector along the straight line, we can write

$$\ddot{\mathbf{r}} = a\mathbf{i}$$

$\Rightarrow \qquad\qquad \dot{\mathbf{r}} = at\mathbf{i} + \mathbf{c},$ where **c** is a vector constant of integration

$t = 0, \dot{\mathbf{r}} = U\mathbf{i}, \Rightarrow \quad \dot{\mathbf{r}} = at\mathbf{i} + U\mathbf{i} \quad$ since $\mathbf{c} = U\mathbf{i}$

$\Rightarrow \qquad\qquad \mathbf{r} = \frac{1}{2}at^2\mathbf{i} + Ut\mathbf{i} + \mathbf{c}'$

$t = 0, \mathbf{r} = \mathbf{0}, \Rightarrow \quad \mathbf{r} = \frac{1}{2}at^2\mathbf{i} + Ut\mathbf{i}.$

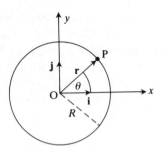

Fig. 1.1

Example 2 Now suppose that the particle moves on a circle radius R at a constant rate. We need to work out what this means. In Fig. 1.1 the axes are a set Oxy for which O is the centre of the circle. θ is the angle between Ox and the radius from O to the particle. Since the particle moves steadily round the circle the rate of change of θ with respect to time is constant, say ω, so that $\dot{\theta} = \omega$, $\ddot{\theta} = 0$.

Hence

$$\mathbf{r} = R\cos\theta\mathbf{i} + R\sin\theta\mathbf{j}$$

$$\dot{\mathbf{r}} = -R\dot{\theta}\sin\theta\mathbf{i} + R\dot{\theta}\cos\theta\mathbf{j}$$

$$|\dot{\mathbf{r}}| = R^2\dot{\theta}^2(\sin^2\theta + \cos^2\theta) = R^2\dot{\theta}^2$$

showing that the velocity has magnitude $R\omega$ and is directed along the tangent since it is perpendicular to $\mathbf{r}$. Differentiating again

$$\ddot{\mathbf{r}} = -R\dot{\theta}^2\cos\theta\mathbf{i} - R\dot{\theta}^2\sin\theta\mathbf{j}$$

since $\ddot{\theta} = 0$. This shows that the acceleration is $R\omega^2$ directed towards the centre of the circle.

Example 3 A charged particle under the influence of a magnetic field can move on a helical path, again at a constant rate. If the axis Oz, having unit vector $\mathbf{k}$ along it, is the central axis of the helix (a helix looks like a spring as used in a spring balance) then the position vector of a point on the helix can be written

$$\mathbf{r} = a\cos\theta\mathbf{i} + a\sin\theta\mathbf{j} + b\theta\mathbf{k}.$$

If the particle moves at a constant rate than again $\dot{\theta} = \omega$, a constant:

$$\dot{\mathbf{r}} = -a\dot{\theta}\sin\theta\mathbf{i} + a\dot{\theta}\cos\theta\mathbf{j} + b\dot{\theta}\mathbf{k}.$$

Using $\ddot{\theta} = 0$

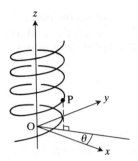

Fig. 1.2

$$\ddot{\mathbf{r}} = -a\dot{\theta}^2 \cos\theta\mathbf{i} - a\dot{\theta}^2 \sin\theta\mathbf{j}.$$

Hence the particle moves with speed $|\dot{\mathbf{r}}| = \sqrt{(a^2 + b^2)}\omega$ along the helix and with acceleration $a\omega^2$ directed towards the central axis of the helix.

1.3   Forces

Newton's second law states that the alteration of motion is produced by the 'motive force impressed'. For example, a stone dropped from a high cliff is said to be acted on by a gravitational force pulling it vertically downwards. The gravitational force is proportional to the mass of the stone, the constant of proportionality being denoted by g, the acceleration due to gravity. The magnitude of a force is measured in newtons (1 newton = 1 kg m s^{-2}) but the force also has a direction. The force on the stone is directed vertically downwards.

A *force* has magnitude and direction and is represented by a vector.

Examples   (a)

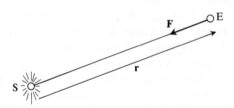

Fig. 1.3

The gravitational force between the Sun and the Earth is

$$\mathbf{F} = -(Gm_s m_e/r^2)\mathbf{e}$$

where **e** is a unit vector parallel to **r**, m_s, m_e are the masses of the Sun and the Earth respectively, and G is the universal gravitational constant ($G = 6.67 \times 10^{-11}$ in SI units).

(b) For a particle at the Earth's surface the gravitational force is mg where m is the mass of the particle, g is the acceleration due to gravity, and

$$|\mathbf{g}| = g = 9.81 \text{ m s}^{-2}$$

acting vertically downwards.

(c) For a moving charged particle in an electromagnetic field the force is

$$\mathbf{F} = e\mathbf{E} + e\mathbf{v} \wedge \mathbf{B}$$

where e is the charge on the particle, **v** is its velocity, **E** is the electric field and **B** is the magnetic induction.

(d) Resistive forces can be experienced by objects moving through a medium such as water or air. In such a case we might expect the force to act in the direction $-\mathbf{v}$ where **v** is the velocity of the particle.

(e) A frictional force resists motion when two surfaces are in contact and in relative motion, e.g. a brick sliding down an inclined plank of wood.

1.4   Newton's second law

We now have the framework in which to express Newton's second law. We make the following definition.

Definition 1.3   An inertial frame of reference is one in which Newton's laws hold.

We will investigate the relationship between inertial frames later in the chapter, but the following examples will give a rough idea. Consider a tennis ball thrown inside a train which is moving with constant speed along a straight track. Experiment shows that the trajectory relative to the train is precisely the same as if the ball had been thrown in the garden. However, if the ball were to be thrown as the train rounds a bend a very different trajectory would be observed. Then it is the case that the train no longer moves uniformly in a straight line, but *accelerates* relative to the Earth's surface. In this last case even over the short time span of the flight of the ball we would not expect a frame fixed in the train to be an inertial frame.

The next step is to express Newton's second law in terms of vector notation, relating force and acceleration. Suppose that a particle of mass m and position vector **r** moves under the influence of forces $\mathbf{F}_i$, $i = 1, 2, \ldots, n$, as shown.

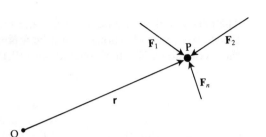

Fig. 1.4

Then the motion of the particle is governed by the following.

Newton's second law:

(NLII) $$m\ddot{\mathbf{r}} = \sum_i \mathbf{F}_i = \mathbf{F}.$$

Mass × acceleration is equal to the vector sum of the forces acting. We can state this more neatly, if we make the following definition.

Definition 1.4   The linear momentum **p** of a particle is its mass × velocity

$$\mathbf{p} = m\dot{\mathbf{r}}.$$

Then the second law can be restated as:

Newton's second law (restated):

The rate of change of linear momentum of a particle is equal to the vector sum of the forces acting on it.

Examples   (a) Suppose that a boy throws a stone in the garden. To set up our system and write down Newton's second law in a mathematical framework we first draw the diagram marking in the forces.

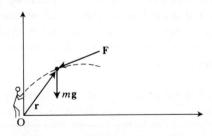

Fig. 1.5

The next step is to choose an origin and mark in the position vector of the stone. A sensible choice would be a point at the boy's feet. We can then write down Newton's second law (NLII) as

$$m\ddot{\mathbf{r}} = m\mathbf{g} + \mathbf{F}$$

where $m\mathbf{g}$ is the gravitational force (which can be regarded as constant so close to the Earth) and $\mathbf{F}$ is the resistive force due to the air.

(b) Consider again the example of the Earth's motion around the Sun. The force on the Earth is $\mathbf{F} = -(Gm_sm_e/r^2)\mathbf{e}$ where $\mathbf{e}$ is a unit vector directed along the line from the Sun to the Earth.

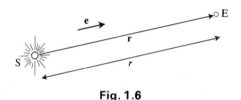

Fig. 1.6

If we choose the centre of the Sun as origin and treat the Earth as a particle with position vector $\mathbf{r}$, then Newton's second law is written as

$$m_e\ddot{\mathbf{r}} = -\frac{Gm_sm_e}{r^2}\mathbf{e}$$

where $\mathbf{e}$ is parallel to $\mathbf{r}$.

1.5   One-dimensional motion

Having written down the equation of motion (NLII) for a particular situation our aim is to solve the equation so that we can calculate the path of the particle *and* make predictions about the behaviour of particles in similar situations. In this section we will restrict ourselves to motion in a straight line, giving a brief outline of the more common soluble examples for those who have not studied mechanics before. The general method of setting up and solving one-dimensional problems is best illustrated by example.

Example 1   A trainee is practising parachute jumping from a high tower, height h from ground level. He opens his parachute immediately and falls from rest in a vertical line to the ground. The information likely to be required is the time taken to fall and the speed on contact with the ground. We have to make some assumption about the force due to air resistance in the parachute. Note that when riding a bicycle into a strong wind the resistance becomes noticeably harder as you increase your speed. The simplest assumption is that the resistance is proportional to the speed of the trainee, say $-\lambda v$. (You might try the same problem with resistance proportional to the square of the speed.) As a crude approximation we shall treat the man as a particle.

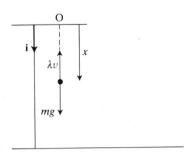

Fig. 1.7

Since the motion is one-dimensional we choose the origin at some point along the straight line in which the particle moves. We have just one direction and hence one axis to consider. In this case we can choose the top of the tower to be the origin and the positive direction along the axis to be vertically downwards. If at each time t we knew the position x of the particle along this axis, then we would have a complete description of the motion. If $\mathbf{i}$ is a unit vector vertically downwards then the position vector of the man is $x\mathbf{i}$. Newton's second law can then be written as

$$m\ddot{x}\mathbf{i} = mg\mathbf{i} - \lambda\dot{x}\mathbf{i}.$$

The vector $\mathbf{i}$ is non-zero and so we have a single scalar equation

$$m\ddot{x} = mg - \lambda\dot{x}.$$

Normally vectors are not introduced at any stage into a one-dimensional problem. We simply write down the scalar equation above.

To solve now becomes a problem in differential equations. At $t = 0$, $x = 0$ and $\dot{x} = 0$ since the man starts from *rest*. The equation is

$$\frac{d^2x}{dt^2} + \frac{\lambda}{m}\frac{dx}{dt} = g.$$

There are several methods of solution. If we solve the homogeneous equation first we get

$$x = A + B \exp[-(\lambda/m)t]$$

and the particular solution is $x = (mg/\lambda)t$, so that adding we have

$$x = A + B \exp[-(\lambda/m)t] + (mg/\lambda)t$$

and using our initial conditions to determine the constants A and B we have

$$x = \frac{m^2g}{\lambda^2}[\exp[-(\lambda/m)t] - 1] + \frac{mg}{\lambda}t.$$

From this we can conclude that the man hits the ground when

$$\frac{m^2g}{\lambda^2}[\exp[-(\lambda/m)t] - 1] + \frac{mg}{\lambda}t = 0.$$

This is an equation in t only which can be solved, using Newton's method for example, if the numerical values of m and λ are known. If the solution of this equation is $t = T$ then the man hits the ground with velocity

$$\dot{x} = (mg/\lambda)\{1 - \exp[-(\lambda/m)T]\}.$$

Notice that the velocity has a maximum value of mg/λ as $t \to \infty$. This is called the terminal velocity.

One point worth mentioning is that we have solved the equation for any values of the constants m, λ and that we can make predictions about the behaviour of parachutists of differing mass! This is important since only numerical work has to be repeated. This is a crucial point and suggests why it is that mathematicians generally work with letters rather than numbers.

Example 2   *Simple harmonic motion.* Our second example comes from an extremely important class of equations. We will look at the simplest practical system which we can conceive. Suppose that an air puck is moving on horizontal glass table attached by a horizontal spring to a fixed point. An air puck operates like a small hovercraft pushing air from its base to minimize friction. If the air puck is set in motion along the line of the taut spring then the motion will take place in that line only and we again have an essentially one-dimensional problem. We assume that the spring satisfies Hooke's law. The spring has a natural length and, if compressed or extended, develops a tension resulting in a force on the air puck. Hooke's law says that the tension is proportional to the extension from the natural length

$$T = \frac{\lambda e}{a} = ke$$

where e is the extension, a is the natural length, and λ is called the modulus. Since both λ and a are constants it is common practice to combine them as above into one *spring constant k*. Hooke's law generally works very well provided the spring does not vary too much from its natural length. The diagram is very simple. Note that in the vertical direction there is no motion so that the normal reaction R on the puck from contact with the table must be equal to the gravitational force mg downwards. We ignore friction assuming that it is very small indeed.

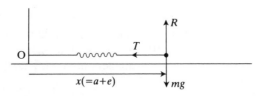

Fig. 1.8

We can choose as origin the fixed end of the spring and measure x, the length along the spring in a positive direction as shown. Then Newton's law (NLII) for the one-dimensional motion of the puck along the spring is

$$m\ddot{x} = -k(x - a)$$

or

$$\ddot{x} + \frac{k}{m}x = \frac{k}{m}a. \tag{1.1}$$

This equation is a standard second-order equation with constant coefficients which has a particular solution $x = a$. This particular solution corresponds to the *equilibrium* solution in which the puck is *stationary*. The homogeneous equation is

$$\ddot{x} + \omega^2 x = 0$$

where $\omega^2 = k/m$. This is called the equation of *simple harmonic motion*. The general solution is

$$x = C \cos \omega t + D \sin \omega t$$

which is often more helpfully written in the equivalent form

$$x = A \cos(\omega t + \phi)$$

where A is called the *amplitude* and ϕ the *phase*. The form of the cosine graph is shown in Fig. 1.9 and clearly indicates that the graph is *periodic*

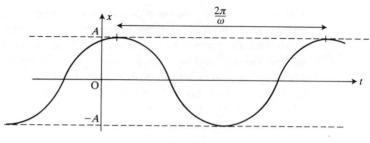

Fig. 1.9

in t with period $2\pi/\omega$. The constant ω is called the (angular) *frequency*. This is the simplest form of periodic oscillatory motion and as such is highly important. In a sense it forms a yardstick against which other oscillatory motions can be compared. The general solution to (1.1) is then

$$x = a + A\cos(\omega t + \phi). \tag{1.2}$$

The amplitude A represents the maximum displacement of the puck from its equilibrium position. This value will be dependent on the initial conditions, the position and velocity of the puck when it is set in motion.

Suppose that at time $t = 0$ the puck is at its equilibrium position and is given a small velocity u in the direction Ox. Then at $t = 0$, $x = a$ and $\dot{x} = u$. When fitting initial values it is generally best to use the alternative form of the general solution. Take

$$x = a + C\cos\omega t + D\sin\omega t$$

$$t = 0, x = a \Rightarrow \quad C = 0$$

$$t = 0, \dot{x} = u \Rightarrow \quad D = u/\omega$$

giving

$$x = a + (u/\omega)\sin\omega t.$$

No matter what initial conditions are used the effect of ignoring friction is to produce a model of 'perpetual motion'. In practice of course the motion decays slowly and finally ceases altogether. In Chapter 3 we will investigate further the circumstances under which a repetitive motion can occur.

Example 3   *Damped vibrations.* When constructing a more realistic model to cover the *damping* caused by frictional forces we think of adding a force due to a *dashpot*. You can think of this schematically as a piston fitting tightly into a cup containing fluid. The result is to produce a resistive force proportional to the relative velocity between the cup and the piston. If this relative velocity is v then the resistive force is written rv where r is the *dashpot constant*.

In example 2 above we could think of the air puck attached to both a spring and dashpot, each of which has O as fixed end. The diagram is essentially unchanged apart from the additional resistive force $F = -r\dot{x}$.

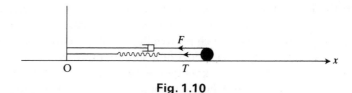

Fig. 1.10

Writing down Newton's law (NLII) as before we have

$$m\ddot{x} = -k(x - a) - r\dot{x}$$

or

$$\ddot{x} + \lambda\dot{x} + \omega^2 x = ka \tag{1.3}$$

where we have set $\lambda = r/m$ and as before $\omega^2 = k/m$.

Notice that the equilibrium position is unchanged by the introduction of a resistive force. As we would expect, $x = a$ is the stationary solution. However, there is a considerable change to the solution of the homogeneous equation

$$\ddot{x} + \lambda\dot{x} + \omega^2 x = 0. \tag{1.4}$$

This is a second-order equation with constant coefficients and has auxilliary equation

$$\beta^2 + \lambda\beta + \omega^2 = 0. \tag{1.5}$$

Remembering that $\lambda > 0$ we have three types of solution given that

$$\beta = \tfrac{1}{2}[-\lambda \pm \sqrt{(\lambda^2 - 4\omega^2)}]. \tag{1.6}$$

(i) $\lambda^2 > 4\omega^2$. There are two real negative roots $-\lambda_1, -\lambda_2$ and solution

$$x = A\exp(-\lambda_1 t) + B\exp(-\lambda_2 t).$$

(ii) $\lambda^2 = 4\omega^2$. There are equal negative root $-\tfrac{1}{2}\lambda$ and solution

$$x = (At + B)\exp[-(\lambda/2)t].$$

(iii) $\lambda^2 < 4\omega^2$. There are two complex roots $-p \pm iq$ where $p = \tfrac{1}{2}\lambda$ and the solution is

$$x = \exp(-pt)(A\cos qt + B\sin qt).$$

In each case the motion is damped but only in the latter case does the oscillatory motion persist, with an exponentially decreasing amplitude. In the first two cases the damping is so strong that it outweighs the undamped oscillation completely. To get the general solution to (1.3) we add the particular solution $x = a$ to the appropriate solution to the homogeneous equation as in (i), (ii), or (iii) above.

Example 4   *Forced oscillations.* The last example of one-dimensional motion that we will consider is that of a particle attached to a spring–dashpot system, one end of which is forced to oscillate. This forms the basis for a simple mathematical model of a seismograph. We suppose that the particle is hanging under gravity attached to one end of the spring, the other end of which is attached to a point P, fixed in a structure which is oscillating in a vertical direction so that the height z of P above the ground satisfies

$$z = h + b \cos \alpha t.$$

We can choose the point at ground level immediately below the particle as origin O with the Ox axis vertically upwards. There are three forces acting on the particle: the tension in the spring, the resistive force, and the gravitational force. The length of the spring is $(z - x)$ where x is the height of the particle above ground level.

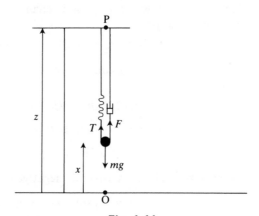

Fig. 1.11

Newton's law (NLII) becomes

$$m\ddot{x} = k(z - x - a) + r(\dot{z} - \dot{x}) - mg$$

or, using $\omega^2 = k/m$ and $\lambda = r/m$,

$$\ddot{x} + \lambda\dot{x} + \omega^2 x = -\omega^2 a - g + \omega^2 z + \lambda\dot{z}. \qquad (1.7)$$

Substituting for z we have

$$\ddot{x} + \lambda\dot{x} + \omega^2 x = -\omega^2 a - g + \omega^2 h + \omega^2 b\cos\alpha t - \lambda\alpha b\sin\alpha t. \quad (1.8)$$

Now provided that there is damping, $\lambda \neq 0$, all solutions to the homogeneous equation decay exponentially, so that the particular solution is the one that prevails in time and is often referred to as the steady state solution. We can find that solution in two parts. First consider the constant term.

$$\ddot{x} + \lambda\dot{x} + \omega^2 x = -\omega^2 a - g + \omega^2 h.$$

Then the particular solution is

$$x = h - a + \frac{g}{\omega^2}.$$

Then we look at the most interesting part of the particular solution. We can express the last two terms on the right-hand side of (1.8) in terms of a complex number so that the equation we are solving becomes

$$\ddot{x} + \lambda x + \omega^2 x = \mathrm{Re}[(\omega^2 b + i\lambda\alpha b)\exp(i\alpha t)].$$

We solve the equation with the complex number on the right and take the real part of x at the end. So we guess a solution $x = C\exp(i\alpha t)$ and calculate the complex number C:

$$-\alpha^2 C + i\lambda\alpha C + \omega^2 C = \omega^2 b + i\lambda\alpha b$$

giving

$$x = \mathrm{Re}\left(\frac{\omega^2 b + i\lambda\alpha b}{\omega^2 - \alpha^2 + i\lambda\alpha}\exp(i\alpha t)\right).$$

Remembering that complex numbers can be written in the form $r\exp(i\theta)$, where r is the modulus and θ is the argument, we may write

$$C = \frac{\omega^2 b + i\lambda\alpha b}{\omega^2 - \alpha^2 + i\lambda\alpha} = A\exp(i\phi).$$

Then

$$x = A\cos(\alpha t + \phi)$$

showing that the amplitude of the forced oscillation is

$$A = \frac{|\omega^2 b + i\lambda\alpha b|}{|\omega^2 - \alpha^2 + i\lambda\alpha|}.$$

The phase ϕ is the argument of C.

The full particular solution can be written

$$x = h - a + \frac{g}{\omega^2} + A\cos(\alpha t + \phi).$$

This is the *steady state* solution.

Corollary to   *Suppose that the system were set up without any resistive force so that* $\lambda = 0$.
example 4   *Then the steady state solution becomes*

$$x = h - a + \frac{g}{\omega^2} + \frac{\omega^2 b}{\omega^2 - \alpha^2}\cos\alpha t$$

giving a perfectly well-behaved solution except as $\alpha \to \omega$. *Here the amplitude becomes larger and larger. This situation is described as* resonance. *Identifying this extreme response is extremely important in the design of structures. Fear of it occurring is said to be the reason for soldiers breaking step when marching across a bridge!*

CONCLUSION

In general there are four parts to the process used when tackling the above examples.

1. Draw a diagram, choosing an origin and marking in forces acting on the particle.

2. Set up the differential equation which governs motion using Newton's second law.

3. Solve the mathematical equation making use of initial conditions if known.

4. Interpret the solution giving the mechanical behaviour of the particle with time.

Of course it may be the case that the equations reached are difficult to solve and that only partial information can be extracted. Then solution and interpretation are developed as fully as possible. The procedure outlined above is substantially the same when determining the mechanical behaviour of a particle in two or three dimensions as we shall see in the next section.

1.6   Particle motion in two- and three-dimensions

The first step is to draw a diagram, identify the forces, and choose a suitable reference point as an origin. We can then write down the vector form of equation of motion (NLII) and set about solving a second-order vector differential equation. This can be a very difficult problem involving choice of a coordinate system, such as Cartesian or polar coordinates. However, in the cases investigated below it is possible to solve the equation and give a complete answer to the question of how the particle moves. If the problem is two-dimensional or more it is often at least as important to know the path of the particle as the time taken to pursue it. Axes and coordinates are chosen and the path found relative to these.

Example 1   Suppose that a stone is thrown with velocity $\mathbf{V}$ at an angle α to the horizontal from a wall height h above the ground. Ignoring air resistance we will calculate the horizontal range of the stone and its path.

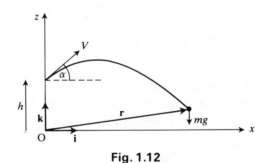

Fig. 1.12

Choosing an origin at ground level immediately below the point of projection and unit vectors along the axes as shown in the diagram, we see that Newton's second law gives

$$m\ddot{\mathbf{r}} = m\mathbf{g}$$

since the gravitational force is the only force acting on the stone.
   Then initially

$$\mathbf{r} = h\mathbf{k} \quad \text{and} \quad \dot{\mathbf{r}} = \mathbf{V} = V\cos\alpha\mathbf{i} + V\sin\alpha\mathbf{k}.$$

Integrating twice and using these conditions at $t = 0$ we have

$$\mathbf{r} = \tfrac{1}{2}t^2\mathbf{g} + t\mathbf{V} + h\mathbf{k}$$

giving

$$\mathbf{r} = -\tfrac{1}{2}gt^2\mathbf{k} + tV\cos\alpha\mathbf{i} + tV\sin\alpha\mathbf{k} + h\mathbf{k} \qquad (1.9)$$

since $\mathbf{g} = -g\mathbf{k}$. This is essentially a two-dimensional path with $\mathbf{r}$ of the form $x\mathbf{i} + z\mathbf{k}$.

The stone hits the ground when $z = \mathbf{r}.\mathbf{k} = 0$, at time t

$$-\tfrac{1}{2}gt^2 + tV\sin\alpha + h = 0$$

with range $x = tV\cos\alpha$, so that the range is

$$\frac{V\cos\alpha}{g}[V\sin\alpha + \sqrt{(V^2\sin^2\alpha + 2gh)}].$$

We can find the path of the particle by setting $\mathbf{r} = x\mathbf{i} + z\mathbf{k}$. Using eqn (1.9) we have

$$x = tV\cos\alpha$$

$$z = -\tfrac{1}{2}gt^2 + tV\sin\alpha + h.$$

Eliminating t

$$z = -\frac{gx^2}{2V^2}\sec^2\alpha + x\tan\alpha + h.$$

Hence the path is a parabola.

Example 2 A ball is thrown with velocity of projection $\mathbf{V}$ in a steady horizontal crosswind of velocity $\mathbf{u}$. Assuming that the air resistance is linearly proportional to velocity and treating the ball as a particle, find the path of the particle parametrically in terms of the time t.

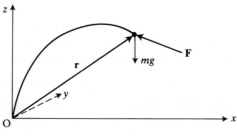

Fig. 1.13

Choose as origin the initial position of the ball. The air resistance $\mathbf{F}$ must be proportional to the *relative velocity* of the ball with respect to the air

$$\mathbf{F} = -\lambda(\dot{\mathbf{r}} - \mathbf{u})$$

where λ is the constant of proportionality. Newton's second law gives

$$m\ddot{\mathbf{r}} = m\mathbf{g} - \lambda(\dot{\mathbf{r}} - \mathbf{u})$$

or

$$\ddot{\mathbf{r}} + \frac{\lambda}{m}\dot{\mathbf{r}} = \mathbf{g} + \frac{\lambda}{m}\mathbf{u}.$$

When solving this type of vector differential equation it pays to keep the vector form for as long as possible, giving greatest simplicity. We solve it by exactly the same method as if it were a scalar equation. First we deal with the homogeneous equation

$$\ddot{\mathbf{r}} + \frac{\lambda}{m}\dot{\mathbf{r}} = \mathbf{0}$$

making a trial solution $\mathbf{r} = \mathbf{B}\exp(\alpha t)$ with $\mathbf{B}$ a constant vector. This gives the auxilliary equation

$$\alpha^2 + \frac{\lambda}{m}\alpha = 0 \quad \Rightarrow \quad \alpha = 0 \text{ or } -\frac{\lambda}{m}.$$

The complementary solution to the homogeneous equation is

$$\mathbf{r} = \mathbf{C} + \mathbf{D}\exp[-(\lambda/m)t].$$

The particular solution is given by guessing that $\mathbf{r} = \mathbf{A}t$ and finding $\mathbf{A}$ by substituting in the inhomogeneous equation to give

$$\mathbf{A} = \frac{m}{\lambda}\mathbf{g} + \mathbf{u}.$$

The general solution to the full equation is then

$$\mathbf{r} = \mathbf{C} + \mathbf{D}\exp[-(\lambda/m)t] + \left(\frac{m}{\lambda}\mathbf{g} + \mathbf{u}\right)t$$

We now use the initial conditions to determine the constant vectors $\mathbf{C}$ and $\mathbf{D}$. At $t = 0$, $\mathbf{r} = \mathbf{0}$ and $\dot{\mathbf{r}} = \mathbf{V}$ giving

$$\mathbf{C} + \mathbf{D} = \mathbf{0}$$

and

$$\mathbf{V} = -\frac{\lambda}{m}\mathbf{D} + \left(\frac{m}{\lambda}\mathbf{g} + \mathbf{u}\right).$$

Finally

$$\mathbf{r} = \frac{m}{\lambda}\left(\mathbf{V} - \frac{m}{\lambda}\mathbf{g} - \mathbf{u}\right)\{1 - \exp[-(\lambda/m)t]\} + \left(\frac{m}{\lambda}\mathbf{g} + \mathbf{u}\right)t.$$

Notice that this gives motion in a plane defined by the two vectors **V** and $(m/\lambda)\mathbf{g} + \mathbf{u}$. The fact that the plane is not vertical is due to the crosswind, of course.

These two examples are essentially simple in a geometrical sense. The interpretation is best done in terms of Cartesian coordinates and axes, two horizontal axes and one vertical. However, in an early example looking at a particle moving on a circular path the position of the particle was investigated using an angle. We shall see in later chapters that choosing a suitable set of coordinates is one of the skills needed in solving mechanical problems.

1.7   Moving frames of reference

The two examples above involve small objects on the surface of the Earth. We have assumed that the Earth's surface provides an inertial frame since we have used Newton's second law to determine the respective motions of the particles. In fact the Earth is rotating about its own axis and about the Sun. We need to know what relationships can exist between two inertial frames, e.g. a frame fixed in the Sun and a frame fixed on the surface of the Earth. Note that over the time span in which the stone or ball was airborne the velocity of the frame of reference used at the surface of the Earth could be taken to be uniform, relative to the centre of the Earth. On the other hand anyone who has tried to walk on a roundabout will know that there is a distinct difference between trying to walk in a straight line relative to the roundabout whilst it is in motion, and doing the same thing once it has ceased turning.

Consider an inertial frame S, origin O and axes Oxyz.

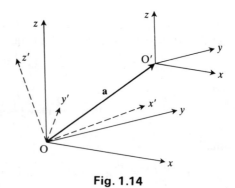

Fig. 1.14

(1) A new set of axes $Ox'y'z'$ fixed in S simply changes the unit vectors used and cannot affect the validity of Newton's laws in the frame.
(2) Similarly a change of origin by a fixed vector $\mathbf{a}$ in S must give another inertial frame.
Of more interest is the following proposition.

Proposition 1.1   *Suppose that* S *is an inertial frame. Then a frame of reference* S', *whose origin* O' *is moving with* uniform *velocity* $\mathbf{U}$ *with respect to* S, *is also an inertial frame.*

Proof

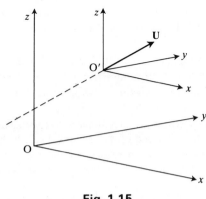

Fig. 1.15

Suppose that the origins O and O' are separated by a vector $\mathbf{a}$ at $t = 0$. Since O' moves with uniform velocity $\mathbf{U}$ with respect to S the vector $\underline{OO'} = \mathbf{a} + \mathbf{U}t$ at a subsequent time t. Then if $\mathbf{r}, \mathbf{r}'$ are the position vectors of a particle with respect to the frames S, S' respectively

$$\mathbf{r}' = \mathbf{r} - (\mathbf{U}t + \mathbf{a})$$

and hence

$$\ddot{\mathbf{r}}' = \ddot{\mathbf{r}}$$

implying that Newton's second law applies in the new frame S', as it applies in S.

The earlier example of a tennis ball thrown inside a train moving with uniform velocity on the Earth's surface is easily seen to fit into the category above. A frame of reference fixed in the train will be an inertial frame relative to an inertial frame fixed on the surface of the Earth. For the time duration of the flight of the tennis ball such a frame on the surface of the Earth exists to a good degree of approximation.

Similarly the motion of an artificial satellite orbiting the Earth can be investigated using a frame with origin at the centre of the Earth. The origin is assumed to be moving with uniform velocity relative to the Sun, to a good degree of approximation, given the time taken by the satellite to complete one orbit. We then use a set of axes, origin the centre of the Earth, which do *not rotate* with the Earth about its polar axis. It is clear that a rotating frame is not an inertial frame with respect to a known inertial frame. The results of this section can be summarized in the following proposition.

Proposition 1.2 *A frame S', O'x'y'z', is an inertial frame relative to a known inertial frame S, Oxyz, if* (i) *O' moves in a straight line with uniform velocity relative to S, and* (ii) *the axes in S' have fixed directions relative to the axes Oxyz of S.*

We might have anticipated (i) by relating it to Newton's first law which states that any massive particle with no forces acting on it moves with uniform velocity in a straight line.

Exercises: Chapter 1

1. A particle mass m is attached to a spring and is hanging vertically downwards. The other end of the spring is fixed. If the spring has spring constant k and natural length a write down the equation of motion of the particle assuming it only moves in a vertical line. What is its equilibrium position?

   If the spring is stretched beyond the equilibrium position by a length $\frac{1}{4}a$ and the particle is released from rest calculate its position at time t. What is the maximum speed achieved by the particle? Write down the period of the motion.

2. Suppose that the situation is as described in the previous question but that the spring is not perfect, in addition acting as a dashpot with constant r. Write down the equation of motion for the particle. What is its equilibrium position?

   If the particle is again set in motion by releasing it from a position of rest $\frac{1}{4}a$ below the equilibrium position calculate its position at time t. If $r^2 < 4mk$ sketch the graph of position against time.

3. A simple pendulum consists of a light inelastic string of length d, fixed at one end O and with the pendulum bob mass m attached to the other end P. The string and bob are free to move in a vertical plane through O and OP makes an angle θ to the downward vertical. Using $\mathbf{i}, \mathbf{j}$ as unit vectors in the plane, $\mathbf{i}$ in the downward vertical direction and $\mathbf{j}$ horizontal, write down a unit vector in the direction of the tension in the string. Hence express Newton's second law in vector form when the string is taut. (See example 2 in Section 1.2, but note that $\dot\theta$ is not constant.) Deduce the equation for the simple pendulum in the form

$$\ddot\theta = -\frac{g}{d}\sin\theta.$$

If θ is small what is the period of small oscillations about the equilibrium position?

Suppose now that the pendulum is no longer confined to a fixed vertical plane through O. If the position vector of the bob from O is $\mathbf{r}$ show that the equation of motion can be written as

$$m\ddot{\mathbf{r}} = mg\mathbf{i} - T\frac{\mathbf{r}}{d}$$

where T is the magnitude of the tension in the string. Show that motion of the bob in a horizontal circle is possible with OP inclined at an angle α to the vertical provided that $T = mg \sec \alpha$ and the angular speed ω of the bob is constant, satisfying $\omega^2 = (g/d) \sec \alpha$.
[Try approaching this problem by setting $\mathbf{r} = d \cos \alpha \mathbf{i} + \mathbf{s}$ where $\mathbf{s}$ is a horizontal vector.]

4. (a) A small stone is thrown in a garden. Prove that its path is in general a parabola (ignoring air resistance). What initial conditions ensure motion in a straight line?
(b) Suppose that we now include air resistance and make the assumption that the magnitude of the resistance is directly proportional to the speed of the stone. Establish the equation of motion and, given that the stone is projected initially with velocity V at an angle α to the horizontal, find the position of the stone at time t, measured from the point of projection. Show directly from your answer that the air resistance reduces the horizontal distance travelled at time t. Show also that as the constant of proportionality tends to 0 the solution $\mathbf{r} = \mathbf{r}(t)$ approaches the solution obtained ignoring resistance.

5. (a) A catapult projects a stone in the normal direction to a playground which slopes at the angle of $10°$ to the horizontal. The initial speed of the stone is 18 m s^{-1}. Ignoring air resistance caclulate the range parallel to the playground.
(b) Suppose that the playground is inclined at an angle β to the horizontal. If the stone were to be fired with velocity V at an acute angle α to the playground, in the vertical plane through the line of greatest slope, find the maximum range parallel to the playground.

6. The three-dimensional analogue of simple harmonic motion has the equation

$$\frac{d^2\mathbf{r}}{dt^2} = -\omega^2\mathbf{r}.$$

Show that the general solution is of the form $\mathbf{r} = \mathbf{A}\cos\omega t + \mathbf{B}\sin\omega t$ where $\mathbf{A}$, $\mathbf{B}$ are arbitrary constant vectors. Hence show that the motion is planar.
(i) Under what conditions on $\mathbf{A}$, $\mathbf{B}$ is the path a circle centred on the origin?
(ii) Suppose that at $t = 0$ $\mathbf{r}$ and $\dot{\mathbf{r}}$ are perpendicular. Show that the path of the particle is an ellipse.
(iii) What conclusions can you draw about the path in the most general form of the solution?

7. Suppose that a (small!) child slides on a rough plane surface inclined at an angle α to the horizontal. In a simplified model of the frictional force during motion we will assume that there is a kinetic coefficient of friction μ such that the magnitude of the friction is $\mu \times$ the magnitude of the normal reaction and directly opposes the velocity of the child across the surface. If $\mathbf{r}$ is the position

vector of the child (treated as a particle!) measured from a fixed origin, establish the equation

$$\frac{d^2\mathbf{r}}{dt^2} = -\mu g \cos \alpha \mathbf{s} + g \sin \alpha \mathbf{i}$$

where $\mathbf{i}$ is a unit vector parallel to the line of greatest slope of the plane and $\mathbf{s}$ is a unit vector whose direction should be specified.

If the surface of the plane is smooth what is the path of the child, assuming the most general initial conditions?

8. The force on a particle, charge e and velocity $\mathbf{v}$, moving under the influence of a constant magnetic field $\mathbf{B}$, is $e\mathbf{v} \wedge \mathbf{B}$. If this is the only force on the particle and if at $t = 0$, $\mathbf{r} = \mathbf{0}$ and $\dot{\mathbf{r}} = \mathbf{V}$, show that

$$m\dot{\mathbf{r}} = e\mathbf{r} \wedge \mathbf{B} + m\mathbf{V}.$$

If a set of coordinate axes is chosen in which $\mathbf{B} = (0, 0, B)$ and $\mathbf{V} = (V_1, 0, V_2)$ show that the path of the particle for $\mathbf{r} = (x, y, z)$ is given by

$$x = \frac{mV_1}{eB} \sin \frac{eBt}{m}$$

$$y = -\frac{mV_1}{eB} + \frac{mV_1}{eB} \cos \frac{eBt}{m}$$

$$z = V_2 t.$$

Sketch the path of the particle.

2 Central forces

2.1 Central forces and angular momentum

In this chapter we consider the application of the Newtonian model to the physical system which originally inspired its formulation – the Sun and its planets. On the basis of experimental evidence we write the attractive force between two gravitating bodies such as the Earth and the Sun, as

$$\frac{GMm}{r^2} \tag{2.1}$$

along the line joining the two bodies, where M, m are the respective masses, r is the distance between them, and G is the universal gravitational constant (6.67×10^{-11} in SI units). Now it turns out that many of the important results obtained from applying Newton's laws to this model also apply to a wider class of forces, the *central forces*. Looking at this enlarged class sheds more light on the properties of physical systems than sticking simply to the inverse square law (2.1) above.

Definition 2.1 A central force $\mathbf{F}$ acting on a particle P depends only on the distance of that particle from some fixed central origin O in an inertial frame and is directed along the line joining the particle to O. If $\mathbf{r}$ is the position vector of the particle from O then

$$\mathbf{F} = F(r)\hat{\mathbf{r}}. \tag{2.2}$$

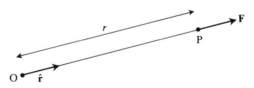

Fig. 2.1

We use $\hat{\mathbf{r}}$ to indicate a vector in the same direction as $\mathbf{r}$ but having unit length.

If a single particle has mass m and is acted on by a central force as above then Newton's second law becomes

$$m\ddot{\mathbf{r}} = F(r)\hat{\mathbf{r}}. \tag{2.3}$$

Considering the path of the Moon around the Earth or of a comet around the Sun it is clear that a feature is the movement of the line joining the body to the central origin, as in Kepler's second law concerning the area swept out by the line as the body follows its path. One important concept linked with this motion is that of *angular momentum.*

Definition 2.2

The angular momentum $\mathbf{L_O}$ of a particle about a point O in an inertial frame S is the moment of linear momentum about O,

$$\mathbf{L_O} = \mathbf{r} \wedge m\mathbf{v} \tag{2.4}$$

where $\mathbf{r}$ is the position vector of the particle *relative* to the point O and $\mathbf{v}$ is the *true* velocity of the particle in the inertial frame.

The definition makes it clear that the angular momentum $\mathbf{L_O}$ depends on the choice of O. In the specific case where the only force acting is a central force we shall see that the angular momentum about the centre of the force is constant.

Suppose that O is some fixed point in the inertial frame. In the special case that $\mathbf{L_O}$ is $\mathbf{0}$ for all times t, then $\mathbf{r}$ and $\mathbf{v}$ are parallel. The particle must move in a straight line with respect to that frame. Conversely if the particle moves in a straight line then the angular momentum about any point on the line is $\mathbf{0}$ at all times. In the general case angular momentum about the origin of a central force is linked to Kepler's second law.

Proposition 2.1

If a particle is acted on solely by a central force with centre O then the angular momentum $\mathbf{L_O}$ is conserved and the path of the particle lies entirely in a fixed plane through O. The motion is planar.

Proof

Taking the vector product of the equation of motion (2.3) with $\mathbf{r}$ gives

$$\mathbf{r} \wedge m\ddot{\mathbf{r}} = \mathbf{r} \wedge F(r)\hat{\mathbf{r}} \tag{2.5}$$

$$= \mathbf{0}.$$

Integrating we have

$$\mathbf{r} \wedge m\dot{\mathbf{r}} = \mathbf{L_O} \tag{2.6}$$

where $\mathbf{L_O}$ is a constant vector. Now taking the scalar product with $\mathbf{r}$ we see that

$$\mathbf{L_O} \cdot \mathbf{r} = 0.$$

Firstly $\mathbf{L_O}$, the angular momentum, is constant and secondly the particle moves in a plane through O perpendicular to $\mathbf{L_O}$. In the special case that $\mathbf{L_O}$ is $\mathbf{0}$ then the particle moves in a straight line through O.

2.2   The scalar equations of motion in plane polar coordinates

Knowing that motion under a central force must take place in a plane through O enables us to choose a sensible set of coordinates. Since the force is a function of distance from O only, we use plane polar coordinates (r, θ) with non-constant unit vectors $\hat{\mathbf{r}}$, $\hat{\boldsymbol{\theta}}$ in the directions of increasing r, θ as shown in Fig. 2.2 below.

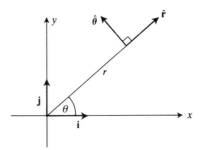

Fig. 2.2

Using column vectors

$$\hat{\mathbf{r}} = \begin{bmatrix} \cos\theta \\ \sin\theta \end{bmatrix} \qquad \hat{\boldsymbol{\theta}} = \begin{bmatrix} -\sin\theta \\ \cos\theta \end{bmatrix}$$

and hence

$$\mathbf{r} = r \begin{bmatrix} \cos\theta \\ \sin\theta \end{bmatrix} = r\hat{\mathbf{r}}.$$

Here $\begin{bmatrix} \cos\theta \\ \sin\theta \end{bmatrix}$ stands for $\cos\theta\,\mathbf{i} + \sin\theta\,\mathbf{j}$.

Differentiating once with respect to time we see that

$$\dot{\mathbf{r}} = \dot{r} \begin{bmatrix} \cos\theta \\ \sin\theta \end{bmatrix} + r\dot{\theta} \begin{bmatrix} -\sin\theta \\ \cos\theta \end{bmatrix}$$

or

$$\dot{\mathbf{r}} = \dot{r}\hat{\mathbf{r}} + r\dot{\theta}\hat{\boldsymbol{\theta}} \tag{2.7}$$

and differentiating again

$$\ddot{\mathbf{r}} = (\ddot{r} - r\dot{\theta}^2)\hat{\mathbf{r}} + (2\dot{r}\dot{\theta} + r\ddot{\theta})\hat{\boldsymbol{\theta}}$$

$$= (\ddot{r} - r\dot{\theta}^2)\hat{\mathbf{r}} + \frac{1}{r}\frac{d}{dt}(r^2\dot{\theta})\hat{\boldsymbol{\theta}}. \tag{2.8}$$

These equations give the components of velocity and acceleration in polar coordinates, which we can now use in the equation of motion (2.3), giving two scalar equations from the vector equation (2.3)

$$m[(\ddot{r} - r\dot{\theta}^2)\hat{\mathbf{r}} + \frac{1}{r}\frac{d}{dt}(r^2\dot{\theta})\hat{\boldsymbol{\theta}}] = F(r)\hat{\mathbf{r}}. \tag{2.9}$$

Here θ is measured from a fixed line in the plane of the motion.

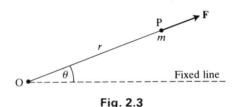

Fig. 2.3

The scalar equations are

$$m(\ddot{r} - r\dot{\theta}^2) = F(r) \tag{2.10}$$

and

$$\frac{m}{r}\frac{d}{dt}(r^2\dot{\theta}) = 0. \tag{2.11}$$

These equations hold in the plane perpendicular to the constant angular momentum vector $\mathbf{L_O}$, which has magnitude $mr^2\dot{\theta}$ in these coordinates, since

$$\mathbf{L_O} = \mathbf{r} \wedge m\dot{\mathbf{r}} = r\hat{\mathbf{r}} \wedge m(\dot{r}\hat{\mathbf{r}} + r\dot{\theta}\hat{\boldsymbol{\theta}}) = mr^2\dot{\theta}\mathbf{n}$$

the unit vector $\mathbf{n}$ being the constant normal to the plane.

Proposition 2.2 (*The scalar equations of motion*) *The magnitude of the angular momentum* $|\mathbf{L_O}|$ *is a constant, mh, where*

$$r^2\dot{\theta} = h \tag{2.12}$$

and

$$m\left(\ddot{r} - \frac{h^2}{r^3}\right) = F(r). \tag{2.13}$$

Proof   Integrate eqn (2.11) with respect to time t to give

$$mr^2\dot{\theta} = mh$$

giving

$$m(\ddot{r} - r\dot{\theta}^2) = m\left(\ddot{r} - \frac{h^2}{r^3}\right) = F(r).$$

These are the scalar equations of motion for a general central force. We hope to be able to solve them in the specific case where the force obeys an inverse square law as the most important application is to gravitating bodies. In particular we should be able to deduce an elliptic orbit for planetary motion as Kepler concluded. Notice also that eqn (2.12) is a statement of Kepler's second law, that is the radius vector to the particle sweeps out the area at a constant rate since area $\delta A \approx \frac{1}{2}r^2 \delta\theta \Rightarrow \dot{A} = \frac{1}{2}r^2\dot{\theta} = \frac{1}{2}h$.

2.3   The equation of path

We could attempt to solve (2.12) and (2.13) finding $r = r(t)$, $\theta = \theta(t)$ in order to find the path of the particle but it generally turns out to be more efficient to find and solve an equation relating r and θ directly, or to be more precise $u = 1/r$ and θ.

Proposition 2.3   *For a particle moving under a central force, setting $u = 1/r$, then*

$$\frac{d^2u}{d\theta^2} + u = -\frac{F(1/u)}{mh^2u^2}. \tag{2.14}$$

Proof   From (2.12) we have $\dot{\theta} = hu^2$, giving

$$\dot{r} = \frac{d}{dt}\left(\frac{1}{u}\right) = -\frac{1}{u^2}\frac{du}{d\theta}\dot{\theta} = -h\frac{du}{d\theta}. \tag{2.15}$$

This is an equation that we shall need again. Differentiating

$$\ddot{r} = \frac{d}{dt}\left(-h\frac{du}{d\theta}\right) = -h\dot{\theta}\frac{d^2u}{d\theta^2} = -h^2u^2\frac{d^2u}{d\theta^2} \tag{2.16}$$

which on substituting in (2.13) gives

$$m\left(-h^2u^2\frac{d^2u}{d\theta^2} - h^2u^3\right) = F\left(\frac{1}{u}\right)$$

and hence the required result.

Before considering examples using the inverse square law, one crucial point concerning *spherical* gravitating bodies should be noted. Provided that the body has spherical symmetry, including a density which depends only on the distance from the centre of the body, then the force on a particle outside the body is exactly the same as if the body were a particle of identical mass situated at its centre.

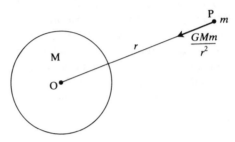

Fig. 2.4

Consequently

$$\mathbf{F}(\mathbf{r}) = -\frac{GMm}{r^2}\hat{\mathbf{r}}$$

where M is the mass of the body, implying that the inverse square law gives a very reasonable approximation to the force (for example) on an artificial satellite orbiting the Earth. (NB The Earth is approximately spherical, but a better approximation can be used if it is taken to be an oblate spheroid, flattened at the poles.)

Example 1   Consider an artificial satellite mass m in orbit around the Earth and at a sufficiently large distance from it to ensure that the Earth's atmosphere does not affect the motion.

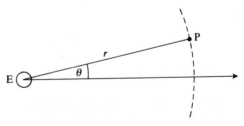

Fig. 2.5

Using proposition 2.3 with $u = 1/r$ we have

$$\frac{d^2 u}{d\theta^2} + u = \frac{GMmu^2}{mh^2u^2}$$

or

$$\frac{d^2 u}{d\theta^2} + u = \frac{\gamma}{h^2}$$

where $\gamma = GM$.

This equation has general solution

$$u = \frac{\gamma}{h^2} + A \cos \theta + B \sin \theta$$

since $u = A \cos \theta + B \sin \theta$ is the general solution to the homogeneous equation

$$\frac{d^2 u}{d\theta^2} + u = 0$$

and $u = \gamma/h^2$ is a particular solution of the inhomogeneous equation.

The general solution can alternatively be written as

$$u = \frac{\gamma}{h^2} + D \cos(\theta + \varepsilon). \tag{2.17}$$

We know that the general polar form of a conic is

$$\frac{c}{r} = cu = 1 + e \cos(\theta + \varepsilon)$$

where the origin ($r = 0$) is situated at one of the foci and the constant ε has been added so that the major axis does not necessarily correspond to the axis $\theta = 0$. (See the Appendix.)

It is now clear that the path of a satellite is a conic and that it will be a closed trajectory according to whether or not the conic is elliptic, given by $0 \le e < 1$. In terms of eqn (2.17) we have $0 \le h^2 D/\gamma < 1$, so that whether the path is an ellipse, a hyperbola, or a parabola depends on its initial velocity and angular velocity and the mass of the attracting body. (Remember $\gamma = GM$.)

Example 2 At a given instant the velocity of a satellite is measured to be $\sqrt{(\gamma/a)}$ at an angle $\pi/4$ to the radius vector and at a distance a from the attracting body. Find the path of the satellite and also the time to complete one loop of the trajectory, known as the period.

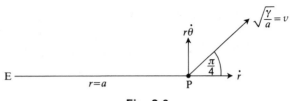

Fig. 2.6

First we must calculate h. Now $h = r^2\dot\theta \Rightarrow h = a\sqrt{(\gamma/2a)} = \sqrt{(a\gamma/2)}$.
Equation (2.14) gives

$$\frac{d^2u}{d\theta^2} + u = \frac{2}{a}$$

$$\Rightarrow \quad u = A\cos\theta + B\sin\theta + \frac{2}{a}.$$

We can *choose* to measure $\theta = 0$ at $t = 0$. Hence when $\theta = 0$, $r = a$,
$\Rightarrow A = -1/a$, and from (2.15)

$$\frac{du}{d\theta} = -\frac{\dot r}{h}\Rightarrow \frac{du}{d\theta} = B = -\sqrt{\frac{\gamma}{2a}}\sqrt{\frac{2}{a\gamma}} = -\frac{1}{a}.$$

Note the use of the equation $du/d\theta = -\dot r/h$. This is always needed when
fitting initial conditions to determine the constants in the equation for the
path.
We get

$$u = -\frac{1}{a}\cos\theta - \frac{1}{a}\sin\theta + \frac{2}{a}$$

$$\Rightarrow \quad u = \frac{2}{a}\left[1 + \frac{1}{\sqrt{2}}\cos\left(\theta + \frac{3\pi}{4}\right)\right].$$

Compare this result with Kepler's first law. This is clearly an elliptic
trajectory with eccentricity $e = 1/\sqrt{2}$.
To find the period T of one complete cycle use the equation $\dot\theta = hu^2$:

$$\dot\theta = \frac{4h}{a^2}\left[1 + \frac{1}{\sqrt{2}}\cos\left(\theta + \frac{3}{4}\pi\right)\right]^2$$

$$\Rightarrow \quad \frac{4}{a^2}\sqrt{\frac{a\gamma}{2}}T = \int_0^{2\pi}\left[1 + \frac{1}{\sqrt{2}}\cos\left(\theta + \frac{3\pi}{4}\right)\right]^{-2}d\theta.$$

Kepler's third law will follow if you persist with the integral!

Example 3   A comet is approaching the Sun from a vast distance with velocity V. If
the Sun exerted no force on the comet it would continue with uniform

velocity V and its distance of closest approach to the Sun would be p. Find the path of the comet and the angle through which it is deflected. Again

$$\frac{d^2u}{d\theta^2} + u = \frac{\gamma}{h^2} \quad \text{where } \gamma = GM_s$$

$$\Rightarrow \quad u = \frac{\gamma}{h^2} + A\cos\theta + B\sin\theta.$$

Notice that the comet starts at a very large distance $R \approx \infty$ and moving at an angle $\approx \pi$ to the radius vector from the centre of the Sun.

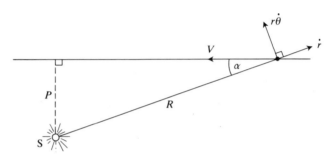

Fig. 2.7

From the diagram we see that

$$\dot{r} = -V\cos\alpha \approx -V$$

and, using $\alpha \approx 0$,

$$h = r^2\dot{\theta} = RV\sin\alpha = pV.$$

At $t = 0$, choose $\theta = 0$ and then $u = 1/R \approx 0$, $du/d\theta = -\dot{r}/h = 1/p$

$$\Rightarrow \quad u = \frac{\gamma}{p^2V^2} - \frac{\gamma}{p^2V^2}\cos\theta + \frac{1}{p}\sin\theta$$

$$\Rightarrow \quad u \to 0 \text{ again when}$$

$$\frac{\gamma}{p^2V^2}\cos\theta - \frac{1}{p}\sin\theta = \frac{\gamma}{p^2V^2}.$$

Solutions to this equation are $\theta = 0$, $2\pi - 2\delta$ where

$$\tan\delta = \frac{pV^2}{\gamma}.$$

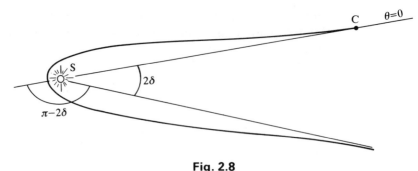

Fig. 2.8

Hence the angle through which the comet is deflected is $\pi - 2\delta$, as shown in the diagram.

2.4   Condition for stability of circular orbits

Many communications satellites are now in everyday use. Some of the most important are those with geostationary orbits, that is those which maintain their position relative to the rotating Earth's surface. If the orbit of a satellite is stationary above a point on the Earth's surface it must have a circular orbit above the equator at a fixed distance from the earth's centre. (Why?)

In general a cicular orbit is possible under the influence of an attractive central force.

Proposition 2.4   (a) *The condition for a circular orbit, $r = a$, under the force $-H(r)\hat{\mathbf{r}}$, is*

$$H(a) = \frac{mh^2}{a^3}.$$ (2.18)

(b) *Further if the orbit is stable then*

$$3H(a) + aH'(a) > 0.$$ (2.19)

Proof   Consider eqn (2.13)

$$m\left(\ddot{r} - \frac{h^2}{r^3}\right) = -H(r).$$

If the orbit is circular then $r = a$, constant, and $\ddot{r} = 0$. Substituting gives the first condition (2.18).

To test stability of the orbit we leave h fixed and replace r by $r = a + \varepsilon$, where ε is regarded as a small parameter depending on t and represents the variation from the circular orbit. If we make the assumption that both $\dot{\varepsilon}$ and $\ddot{\varepsilon}$ are also small, substitution in (2.13) gives

$$m\left(\ddot{\varepsilon} - \frac{h^2}{(a + \varepsilon)^3}\right) = -H(a + \varepsilon).$$

Expanding to first order in ε and using Taylor's theorem we see that

$$m\left[\ddot{\varepsilon} - \frac{h^2}{a^3}\left(1 - \frac{3\varepsilon}{a}\right)\right] = -H(a) - \varepsilon H'(a) + O(\varepsilon^2)$$

$$m\ddot{\varepsilon} + \left(\frac{3h^2m}{a^4} + H'(a)\right)\varepsilon = O(\varepsilon^2)$$

where $O(\varepsilon^2)$ means that the terms ignored are at least as small as ε^2.

Compare this with the equation $\ddot{\varepsilon} + c\varepsilon = 0$. Then for $c > 0$ we may write $c = \omega^2$ and the resulting equation is the standard equation for simple harmonic motion. If on the other hand $c < 0$ then the equation has exponential solutions $\exp(\pm pt)$, setting $c = -p^2$. One of these solutions has exponential growth and indicates that ε can grow very large. If $c = 0$ the solution for ε is linear and again the solution can become large. Only the case $c > 0$ ensures a periodic variation in ε for all t.

Looking at our approximate equation for ε above this means that we must have

$$\left(\frac{3h^2m}{a^4} + H'(a)\right) > 0$$

which gives

$$3H(a) + aH'(a) > 0$$

on using (2.18) the condition for a circular orbit. Of course this only guarantees stability correct to a first-order approximation. If possible a full solution of the original equation with appropriate initial conditions will guarantee complete stability.

Corollary *A satellite orbiting the Earth in a geostationary orbit is above the equator at a distance of 42 000 km from the centre of the Earth. The orbit is stable.*

Proof In order that the satellite stays above the same point on the Earth's rotating surface the only plane through the centre of the Earth which can contain its path is the plane through the equator. The satellite has the same angular velocity as the Earth about its polar axis. Since $h = r^2\dot{\theta}$ and $\dot{\theta}$ is constant, the orbit must be circular, r being constant. Using (2.18) we have

$$r^2\dot\theta = h \quad \text{and} \quad \frac{h^2}{r^3} = \frac{GM}{r^2}.$$

The angular velocity of the Earth is 2π radians per day which can be converted to radians per second. The universal gravitational constant $G = 6.67 \times 10^{-11}$ in SI units and the mass of the Earth is 5.98×10^{24} kg. Substituting these values we find that the radius of the circular orbit is approximately 42 000 km.

For the inverse square law force we have $H(r) = \gamma/r^2$, giving

$$3H(a) + aH'(a) = \gamma/a^2$$

and the stability condition must hold since $\gamma > 0$. This has most important practical consequences since any communications satellite would rapidly disappear if the orbit were not stable under small deviations from its circular path!

We can also show that the orbit is stable if the magnitude of the angular momentum mh is increased and also if the direction of the normal to the plane of the motion is varied by a small angle. Each of these alterations to the path simply changes the orbit to an ellipse close to the original circle.

2.5   General examples

We will look at two examples in which the force involved is not given by the inverse square law.

Example 4   Suppose that a particle mass m with position vector $\mathbf{r}$ in an inertial frame is acted on by a force $-m\omega^2\mathbf{r}$. The vector equation of motion is

$$m\ddot{\mathbf{r}} = -m\omega^2\mathbf{r}.$$

The motion must be planar as this is a central force. The general solution to the equation is

$$\mathbf{r} = \mathbf{A}\cos\omega t + \mathbf{B}\sin\omega t$$

where $\mathbf{A}$ and $\mathbf{B}$ are arbitrary constant vectors which could be found if initial conditions are known. In general the path of the particle will be an ellipse. Unlike the elliptic path found when an inverse square force applies, the centre of this ellipse is the centre of the force. (See exercise 6, Chapter 1.)

Example 5   Suppose that a particle mass m is acted on by a force $\alpha r^{-2} + \beta r^{-3}$ per unit mass (where $\beta = \frac{1}{2}\alpha a$) directed towards the origin $r = 0$ of an inertial frame. At time $t = 0$ measurements of distance and velocity of the particle show that it is a distance a from the origin moving with velocity $\sqrt{(\alpha/a)}$ in a direction perpendicular to the radius vector. Find the maximum and minimum distances of the particle from the origin and the angle turned through by the radius vector in traversing between them.

The equation of path (2.14) gives

$$\frac{d^2u}{d\theta^2} + u = \frac{\alpha u^2 + \beta u^3}{h^2 u^2}$$

so that

$$\frac{d^2u}{d\theta^2} + \left(1 - \frac{\beta}{h^2}\right)u = \frac{\alpha}{h^2}. \tag{2.20}$$

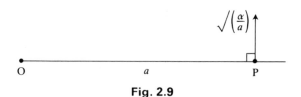

Fig. 2.9

Choosing $\theta = 0$ at $t = 0$ we have $r = a$, $\dot{r} = 0$, and $r\dot{\theta} = \sqrt{(\alpha/a)}$. Hence $h = \sqrt{(\alpha a)}$ and (2.20) becomes

$$\frac{d^2u}{d\theta^2} + \frac{1}{2}u = \frac{1}{a} \quad \text{since } \beta = \frac{1}{2}\alpha a.$$

Solving we get

$$u = \frac{2}{a} + A\cos\frac{\theta}{\sqrt{2}} + B\sin\frac{\theta}{\sqrt{2}}.$$

Putting in the initial conditions, and remembering that $du/d\theta = -\dot{r}/h$, gives

$$u = \frac{2}{a} - \frac{1}{a}\cos\frac{\theta}{\sqrt{2}}.$$

Minimum r corresponds to maximum u and vice versa, so that the maximum value of r is a when $\theta = 0$ and the minimum value of r is $a/3$ when $\theta = \sqrt{(2)}\pi$. The angle turned through must be $\sqrt{(2)}\pi$.

In the next chapter we will see that by using energy considerations we can give an account of the circumstances under which closed paths can exist, a subject of vital importance for life on this planet.

Exercises:
Chapter 2

1. Derive the equation

$$\frac{d^2u}{d\theta^2} + u = \frac{GM}{h^2}$$

for a satellite moving in the Earth's gravitational field.

A spacecraft is orbiting the Earth in a circular orbit. A sudden rocket burst increases its speed to V (assumed along the tangent to the circle which has radius R from the centre of the Earth). Find the conditions that the spacecraft (a) continues to orbit the Earth, (b) escapes. Describe the resulting orbits.

If $R = 10\,000$ km what is the minimum escape velocity?
[The universal gravitational constant $G = 6.67 \times 10^{-11}$ in SI units and the mass of the earth $M = 5.98 \times 10^{24}$ kg.]

2. A particle is attracted towards a fixed point O by a force of magnitude $(\alpha r + \beta^2)/r^3$ per unit mass. Show that the motion will take place in a plane. By means of a suitable choice of polar coordinates in the plane and the substitution $u = 1/r$ show that the orbital equation can be written in the form

$$\frac{d^2u}{d\theta^2} + u = \frac{\alpha}{h^2} + \frac{\beta^2}{h^2}u.$$

Initially the particle is at a distance $\beta^2/(3\alpha)$ from O and is moving with a speed of $4\alpha/\beta$ in a direction making an angle of $\pi/3$ with the radius vector pointing towards O. Show that it returns to its initial position after one revolution around O and then flies off to infinity. What is the angle between the initial and final velocities of the particle?

(Oxford)

3. A particle is *attracted* to a fixed point O by a force $F(r)$ per unit mass, where r is the distance of the particle from O. Assuming that the particle moves in a plane through O prove that

$$\frac{d^2u}{d\theta^2} + u = \frac{F(u^{-1})}{h^2u^2}$$

where h is a constant, $u = r^{-1}$, and θ is the angle the line from O to the particle makes with some fixed direction in the plane.

Now suppose that

$$F(r) = k\left[4\left(\frac{a}{r}\right)^2 - 3\left(\frac{a}{r}\right)^3\right]$$

and the particle is projected from a point at a distance a from O with a component of velocity $\sqrt{(ak)}$ along the line from O to the particle extended and an equal component perpendicular to the line. Show that the maximum and minimum distances of the particle from O are $2a$ and $\frac{2}{3}a$ respectively. Find the angle turned through by the radius vector between the directions corresponding to the first maximum and subsequent minimum.

(Oxford)

4. A comet is tracked as it passes through the solar system. It is observed that the distance of closest approach to the Sun is d and at that point it has speed V. What is the minimum value of V? Ignoring any effect of the planets what is the condition that the comet is a recurrent visitor to the solar system? In the case that the condition holds find the maximum distance of the comet from the Sun.

5. A particle P is attracted towards a fixed point O in an inertial frame by a force λr^{-3} per unit mass where $r = OP$. The particle is set in motion at a distance a from O with speed v perpendicular to OP.
   (i) Show that circular motion occurs if $\lambda = a^2 v^2$.
   (ii) If $\lambda > a^2 v^2$ show that $u = r^{-1} = a^{-1} \cosh \omega\theta$ where ω should be determined, and that the particle reaches O in time $a^2 / \sqrt{(\lambda - a^2 v^2)}$.
   (iii) If $\lambda < a^2 v^2$ show that r increases indefintely.

   (Oxford)

6. Two stars P_1, P_2 (with mass m_1, m_2 respectively) are isolated so that the only force on each is the force due to the other. If in an inertial frame with origin O the position vectors of P_1 and P_2 are r_1 and r_2, write down Newton's second law for each particle. Show that

$$\ddot{r} = -\frac{G(m_1 + m_2)}{r^2} \hat{r}$$

where $r = r_2 - r_1$ and $\hat{r}$ is a unit vector in the same direction as r. Deduce that the line $P_1 P_2$ is always perpendicular to a fixed vector $h = r \wedge \dot{r}$ in the inertial frame.

   If the line $P_1 P_2$ rotates at right angles to h with constant angular speed ω show that the two stars remain at a fixed distance R apart provided that

$$R^3 \omega^2 = G(m_1 + m_2).$$

Show that this motion is stable with respect to a small variation in their distance apart.
[Hint: consider $P_1 P_2$ in a plane perpendicular to h and use plane polar coordinates as for a single particle.]

7. An artificial satellite is moving under the gravitational influence of the Earth and also under the resistance of the Earth's outer atmosphere. Suppose that the resistance is modelled as directly opposite to the velocity of the particle (antiparallel), but not necessarily proportional in magnitude. Regarding the centre of the Earth as the origin of an inertial frame to a good degree of approximation write down Newton's second law for the satellite. From your equation give an argument as to whether or not the motion is planar.

3 Energy

3.1 The one-dimensional case

When a ski-jumper starts his run down the hill he has very little initial velocity. As he loses height so he gains speed. Conversely in order to gain sufficient height to reach a fire in a tall building the nozzle velocity of water from a fireman's hose has to be of a high order of magnitude. There is some balancing mechanism between the work done overcoming the force of gravity and the change in the velocity of the particle.

Some readers will be familiar with the one-dimensional formalism of energy. Newton's second law leads us to a very clear description which we recapitulate here.

Suppose a particle, mass m, moving along a straight line, is acted on by a total force F which depends only on the position, x, of the particle. Then Newton's second law can be written

$$m\ddot{x} = F(x). \tag{3.1}$$

The energy which the particle posesses by virtue of its motion is called *kinetic energy*, being, by definition, $\frac{1}{2}m\dot{x}^2$. The particle also has energy by virtue of the total force acting on it. This is called the *potential energy*.

Definition 3.1   The potential energy $V(x)$, arising from a total force $F(x)$, is

$$V(x) = -\int F(x)\mathrm{d}x.$$

We can see that the potential energy is defined up to an additive constant so that its zero level has to be chosen for each individual example. This definition can equivalently be written

$$F(x) = -\frac{\mathrm{d}V}{\mathrm{d}x}.$$

Proposition 3.1   *If the total force acting on the particle is derived from a potential energy then the total energy is* conserved.

$$\tfrac{1}{2}m\dot{x}^2 + V(x) = E \tag{3.2}$$

where E is the constant total energy.

Proof
Writing $F(x) = -dV/dx$ and multiplying eqn (3.1) by $\dot{x}$ we have

$$m\ddot{x}\dot{x} = F(x)\dot{x}$$

and integrating with respect to t

$$\tfrac{1}{2}m\dot{x}^2 = \int F(x)\dot{x}\,dt + \text{constant}$$

$$= \int F(x)\,dx + \text{constant}.$$

Hence

$$\tfrac{1}{2}m\dot{x}^2 + V(x) = E$$

where E is the constant total energy.

This proposition enables us to solve a first-order differential equation with a constant E in place of the previous second-order equation. This can often produce a simplificiation. In those cases where the analytic solution of the problem is not known the energy equation can be extremely useful. The total energy E is fixed if initial conditions are given, involving known values for x and $\dot{x}$ at time $t = t_0$.

In particular, if the particle is moving vertically under gravity, x being measured vertically upwards, the force on the particle is $F = -mg$ with corresponding potential energy

$$V(x) = mgx$$

choosing $x = 0$ as the base line from which to measure the potential energy. We have

$$\tfrac{1}{2}m\dot{x}^2 + mgx = E$$

constant total energy. This equation can be interpreted as meaning: *loss in kinetic energy equals gain in potential energy and vice versa.* This expression accounts for the trade-off between height and speed in the cases of the ski-jumper and the water from the fire hose.

Notice that the stipulation required to deduce conservation of energy in the one-dimensional case is that the total force acting on the particle is a function of position only.

Example 1
Consider a particle attached to one end of a vertical spring, the tension in the spring satisfying Hooke's law. The other end of the spring is attached to a fixed point in the laboratory. Suppose that the spring constant is k, its natural length is a, and the mass of the particle is m. We

can choose as origin the fixed end of the spring and measure the position x of the particle vertically downwards. The next move is to draw the diagram marking in the forces.

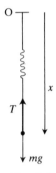

Fig. 3.1

Newton's second law becomes

$$m\ddot{x} = mg - k(x - a). \tag{3.3}$$

Now $V(x) = -\int F(x)\,dx$ and as integration is additive we can consider the two forces separately.

1. The potential energy due to gravity can be taken to be $-mgx$. Note that the sign has changed. This is due to x being measured vertically *downwards*. Also we have elected to take the zero of this potential to be at the fixing point of the spring.

2. The potential energy stored in the spring is

$$-\int -k(x - a)\,dx = \tfrac{1}{2}k(x - a)^2 + \text{constant}.$$

It is *usual* to take the constant to be zero so that the *potential energy stored in the unstretched spring is zero*.

The total potential energy is given by the sum of these two:

$$V(x) = -mgx + \tfrac{1}{2}k(x - a)^2. \tag{3.4}$$

Proposition 3.1 tell us that the total energy is conserved and we have

$$\tfrac{1}{2}m\dot{x}^2 - mgx + \tfrac{1}{2}k(x - a)^2 = E. \tag{3.5}$$

Example 2   In a similar example consider the simpler case of the air puck whose motion is horizontal. If we ignore friction then as in the first chapter Newton's second law becomes

$$m\ddot{x} = -k(x - a).$$

Then the energy equation (3.2), using (2) above, becomes

$$\tfrac{1}{2}m\dot{x}^2 + \tfrac{1}{2}k(x - a)^2 = E. \tag{3.6}$$

Since both terms on the right-hand side are positive and E is constant, this equation ensures that the motion executed by the air puck is periodic. It places maximum values on both $|\dot{x}|$ and $|x - a|$, since $\dot{x}^2 \leq 2E/m$ and $(x - a)^2 \leq 2E/k$. This is what we would expect since we have already shown that the general solution is harmonic about $x = a$.

On the other hand recall that a general solution for the same equation with damping added (no forcing term and so homogeneous) is a solution which *decays* to an equilibrium position in all three types of solution. These simple examples suggest that a periodic motion is possible only if the total energy is conserved.

3.2   The general case

We start from Newton' second law in three dimensions

$$m\ddot{\mathbf{r}} = \mathbf{F}$$

where $\mathbf{F}$ is the vector sum of all the forces acting on the particle. It seems clear that if we are looking for an equivalent statement about energy, the force $\mathbf{F}$ can only be a function of $\mathbf{r}$, the position vector of the particle. The kinetic energy is a scalar (see below) and so an energy conservation equation would be a scalar first-order differential equation obtained from a second-order vector differential equation. The situation is more complicated than the relatively simple one-dimensional problem.

Consideration of gravitational examples again gives us a clue as to how to extend these ideas to three-dimensional motion. Consider a man pushing a wheelbarrow filled with bricks. Pushing the barrow on level ground is relatively easy, but to get a load up a ramp and into a skip he takes a run at it increasing his kinetic energy at the bottom of the ramp. He has to work much harder to push the load up the ramp than to push it the same distance on the level. What is important here is the increase in height, that is the change in position which is in the *same direction* as the gravitational force, the vertical direction. We think intuitively of energy as the amount of effort or work required to perform a given task.

This leads us to the following definition.

Definition 3.2

If a force $\mathbf{F}$ moves its point of application by a vector $\delta\mathbf{r}$ of small magnitude then we say the *work done* by the force is

$$\delta W = \mathbf{F}.\delta\mathbf{r}. \tag{3.7}$$

Hence the work done is the component of the force in the direction of $\delta\mathbf{r}$ multiplied by the distance moved, the magnitude of $\delta\mathbf{r}$. In one dimension (3.7) reduces to

$$\delta W = F\delta x = -\delta V.$$

For those students unfamiliar with integration of vectors the following paragraphs in smaller type can be ignored. They are included for completeness of the argument and are not essential in order to understand the main examples included in this chapter.

If the point of application moves along a curve C from point A to point B then the total work done is the line integral

$$W = \int_A^B \mathbf{F}.d\mathbf{r} \tag{3.8}$$

taken along the curve C. If the curve C is given parametrically by $\mathbf{r} = \mathbf{r}(u)$ and $\mathbf{F} = \mathbf{F}(\mathbf{r})$ then

$$W = \int_A^B \mathbf{F}.\frac{d\mathbf{r}}{du}\,du.$$

Equation (3.8) looks very similar to our definition of $V(x)$ in the one-dimensional case, if we were to take $V = -W$. We could try to define V in this way. However the line integral (3.8) depends in general not only on the endpoints A, B but also on the path taken. We settle for considering only those forces for which the definition works.

Definition 3.3

A force $\mathbf{F}$ is said to be conservative if the total work done, when its point of application traverses once any closed curve C in its domain, is zero,

$$\int_C \mathbf{F}.d\mathbf{r} = 0. \tag{3.9}$$

Examples

(a) Any constant force $\mathbf{F} = \mathbf{F}_0$ is conservative since

$$\int_C \mathbf{F}_0{\cdot}d\mathbf{r} = \Delta[\mathbf{F}_0.\mathbf{r}] = 0$$

where Δ means the change in the function which occurs on traversing the curve C once.

(b) The gravitational force $F = -(GMm/r^3)r$, $r \neq 0$, is conservative, since

$$\int_C -(GMm/r^3)r.dr = \int_C -(GMm/r^2)dr = \Delta[GMm/r] = 0$$

for $r \neq 0$, since $r.dr = d(\frac{1}{2}r^2) = d(\frac{1}{2}r^2) = rdr$.

(c) Suppose that $F = yi - xj + zk$ and let C by the circle

$$x^2 + y^2 = a^2 \qquad z = b.$$

Using the usual parametrization $x = a\cos\theta$, $y = a\sin\theta$, then

$$\int_C F.dr = \int_0^{2\pi} (a\sin\theta i - a\cos\theta j + bk).(-a\sin\theta i + a\cos\theta j) \, d\theta$$

$$= -2\pi a^2.$$

Clearly F is not conservative.

We need to know whether or not a force is conservative and a corresponding definition of associated potential.

Proposition 3.2   *The following conditions are equivalent:*
(i) $\int_C F.dr = 0$ *for all closed loops* C,
(ii) $\int_A^B F.dr$ *is independent of the path taken from* A *to* B,
(iii) $\operatorname{curl} F = 0$, *everywhere in its domain,*
(iv) *there exists a scalar function* ϕ *such that*

$$F = -\operatorname{grad} \phi.$$

Proof   (i) $\Leftrightarrow$ (ii). Any two paths from A to B form a closed loop if one of the paths is traversed in reverse direction.
(i) $\Rightarrow$ (iii). This follows from Stoke's theorem which states that

$$\iint_S \operatorname{curl} F.dS = \int_C F.dr$$

for the closed loop C bounding the surface S. Since this holds for any surface S with boundary curve C then we conclude that $\operatorname{curl} F$ is zero everywhere.
(iii) $\Rightarrow$ (iv). It is a well-known result from vector analysis that if $\operatorname{curl} F \equiv 0$ then such a scalar function ϕ exists.

(iv) $\Rightarrow$ (ii).
$$\int_A^B F.dr = \int_A^B -\operatorname{grad} \phi.dr = -\int_A^B d\phi$$

$$= -\phi(B) + \phi(A)$$

which shows that the integral depends only on the endpoints. This completes the proof.

We can now define what we mean by the potential associated with a force.

Definition 3.4

If **F** is a conservative force then the scalar function $\phi(\mathbf{r})$ such that

$$\mathbf{F} = -\operatorname{grad} \phi \tag{3.10}$$

is called the potential associated with the force, ϕ being uniquely determined up to an additive constant. Note that if grad ϕ exists then **F** must be conservative.

Definition 3.5

If ϕ is a scalar function of position $\mathbf{r} = x\mathbf{i} + y\mathbf{j} + z\mathbf{k}$, grad ϕ is defined to be

$$\operatorname{grad} \phi = \frac{\partial \phi}{\partial x}\mathbf{i} + \frac{\partial \phi}{\partial y}\mathbf{j} + \frac{\partial \phi}{\partial z}\mathbf{k}. \tag{3.11}$$

Then grad ϕ is a vector-valued function.

In one dimension only these definitions reduce to precisely the definition of potential energy in Section 3.1, since $\mathbf{F} = -(d\phi/dx)\mathbf{i}$ if $\phi = \phi(x)$.

Examples

(a) $\mathbf{F} = -mg\mathbf{k}$, then $\phi = mg\mathbf{k}.\mathbf{r} = mgz$, **k** being a unit vector measured vertically upwards
(b) $\mathbf{F} = -(GMm/r^2)\hat{\mathbf{r}}$, then $\phi = -(GMm/r)$. To see this set $r = \sqrt{(x^2 + y^2 + z^2)}$.

These two examples of the potential associated with a single force are important ones to consider. They represent the forms of the potential energy associated with the constant gravitational force at the surface of the Earth and the potential energy associated with the force between two astronomical bodies.

[(c) $\mathbf{F} = y\mathbf{i} - x\mathbf{j} + z\mathbf{k}$ gives

$$\operatorname{curl} \mathbf{F} = \det \begin{bmatrix} \mathbf{i} & \mathbf{j} & \mathbf{k} \\ \dfrac{\partial}{\partial x} & \dfrac{\partial}{\partial y} & \dfrac{\partial}{\partial z} \\ y & -x & z \end{bmatrix}$$

$$\operatorname{curl} \mathbf{F} = -2\mathbf{k}.$$

Thus no potential can exist for this force.]

For a single conservative force acting on a particle the work done δW is given by

$$\delta W = \mathbf{F}.\delta \mathbf{r}$$

$$= - \operatorname{grad} \phi.\delta \mathbf{r}$$

$$= - \left(\frac{\partial \phi}{\partial x} \delta x + \frac{\partial \phi}{\partial y} \delta y + \frac{\partial \phi}{\partial z} \delta z \right).$$

Hence by the chain rule for partial differentiation we have that

$$\delta W = - \delta \phi. \tag{3.12}$$

For a single force acting on a particle for which we have $\mathbf{F} = - \operatorname{grad} \phi$ then the work done in a small change of position is equal and opposite to the change in the potential.

3.3   Conservation of energy

We can now look at the general case of a particle moving in an inertial frame, origin O, position vector $\mathbf{r}$, under total force $\mathbf{F} = \mathbf{F}(\mathbf{r})$. By Newton's second law

$$m\ddot{\mathbf{r}} = \mathbf{F}. \tag{3.13}$$

As we have already suggested the particle has energy by virtue of its motion.

Definition 3.6 A particle, mass m, velocity $\mathbf{v}$, has *kinetic energy* $\frac{1}{2}m\mathbf{v}^2$.

This definition gives kinetic energy a scalar value involving as it does the square of the speed $|\mathbf{v}|$ $(|\mathbf{v}|^2 = \mathbf{v}^2)$.

With the definition of kinetic energy above we shall be able to deduce that total energy is conserved if we can also define a potential energy.

Definition 3.7 If the total force acting on the particle is $\mathbf{F} = \Sigma_i \mathbf{F}_i$ where each force $\mathbf{F}_i$ is conservative and derived from a potential ϕ_i then V, the potential energy of the particle, is

$$V = \sum_i \phi_i. \tag{3.14}$$

Proposition 3.3 *It follows immediately from this definition that*

$$V = - \int \mathbf{F}.d\mathbf{r} \Leftrightarrow \delta V = - \mathbf{F}.\delta \mathbf{r}$$

and

$$\mathbf{F} = -\operatorname{grad} V \qquad (3.15)$$

where **F** is the total force on the particle and is conservative.

Proposition 3.4 *If those forces which do work in the motion are conservative then energy is conserved and*

$$\tfrac{1}{2}m\dot{\mathbf{r}}^2 + V(\mathbf{r}) = E, \quad \text{constant.} \qquad (3.16)$$

Proof Taking the scalar product of the Newtonian equation of motion with $\dot{\mathbf{r}}$

$$m\ddot{\mathbf{r}}.\dot{\mathbf{r}} = \mathbf{F}.\dot{\mathbf{r}}$$

and integrating with respect to t we have

$$\tfrac{1}{2}m\dot{\mathbf{r}}^2 = \int \mathbf{F}.\dot{\mathbf{r}}\,dt + \text{constant.}$$

Since **F** is conservative the integral on the right is independent of the path, so that the right-hand side varies solely with the position of the particle.
Then

$$\int \mathbf{F}.\dot{\mathbf{r}}\,dt = \int \mathbf{F}.d\mathbf{r}$$

$$= -V(\mathbf{r}) + \text{constant}$$

and the proposition is proved.

We have deduced that there is conservation of energy if the total force acting on the particle is conservative. Note that if there are forces which contribute no work in the subsequent motion of the particle and the remaining forces are conservative then energy will still be conserved. For example, if a small particle is moving on the surface of a smooth bowl then the normal reaction does no work and the force of gravity is conservative. Total energy is conserved.

Examples (a) Consider a small air puck sliding on a smooth glass plate which is tilted at an angle α to the horizontal.

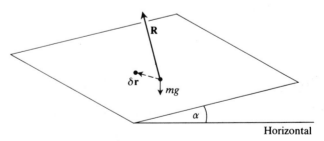

Fig. 3.2

Choose an origin O at a suitable fixed point on the plate. For any change $\delta\mathbf{r}$ in the position of the puck on the surface of the plane, the normal reaction, $\mathbf{R}$, does no work, since $\mathbf{R}.\delta\mathbf{r} = 0$. We have

$$\tfrac{1}{2}m\dot{\mathbf{r}}^2 + mg\mathbf{k}.\mathbf{r} = E$$

where

$$V = mg\mathbf{k}.\mathbf{r}.$$

In any projectile motion for which air resistance is negligible we have shown that energy is conserved. However:

(b) For a projectile moving under gravity with air resistance linearly dependent on velocity Newton's second law can be written

$$m\ddot{\mathbf{r}} = -mg\mathbf{k} - \lambda\dot{\mathbf{r}}$$

where $\mathbf{r}$ is the position of the particle. Suppose that the projectile moves on a closed path $\mathbf{r} = \mathbf{r}(t)$. Then the work done by the resistive force in small time δt is

$$\delta W = -\lambda\dot{\mathbf{r}}.\delta\mathbf{r} = -\lambda\dot{\mathbf{r}}.\dot{\mathbf{r}}\delta t < 0$$

$$\Rightarrow \quad -\lambda \int \dot{\mathbf{r}}.\dot{\mathbf{r}}\,dt < 0 \text{ around a closed path.}$$

Since $\delta W < 0$ no matter which direction the particle moves in, air resistance must be a *dissipative* force and energy is lost. Similarly any frictional force *dissipates* energy from a mechanical system.

A single particle moving in a conservative system of forces may perform a particularly important type of motion. Suppose that eqn (3.16) holds. Then

$$\tfrac{1}{2}m\dot{\mathbf{r}}^2 + V(\mathbf{r}) = E.$$

If from some $\mathbf{r}_0$ the particle returns at a subsequent time to that same position $\mathbf{r}_0$ it must do so with the same kinetic energy and hence the same speed. It follows that in a conservative system it is possible for *closed trajectories* to occur. This is a crucial point when considering the motion of our own Earth around the Sun.

3.4   Application to central forces

First of all we show that all central forces are conservative.

Proposition 3.5   *The central force* $\mathbf{F}(\mathbf{r}) = F(r)\hat{\mathbf{r}}$ *is conservative with potential* ϕ *such that*

$$\phi(r) = -\int F(r)\,dr. \tag{3.17}$$

Proof   Using $Oxyz$, a system of Cartesian axes at O, the centre of the force, we can write the position vector as $\mathbf{r} = x\mathbf{i} + y\mathbf{j} + z\mathbf{k}$ and $r = \sqrt{(x^2 + y^2 + z^2)}$. If $\phi = \phi(r)$ only

$$\nabla\phi = \frac{\partial\phi}{\partial x}\mathbf{i} + \frac{\partial\phi}{\partial y}\mathbf{j} + \frac{\partial\phi}{\partial z}\mathbf{k}$$

$$= \phi'(r)\left(\frac{x}{r}\mathbf{i} + \frac{y}{r}\mathbf{j} + \frac{z}{r}\mathbf{k}\right)$$

so that

$$\nabla\phi = \phi'(r)\hat{\mathbf{r}}.$$

Hence if $\mathbf{F}(r) = -\nabla\phi$ then

$$\phi'(r) = -F(r)$$

and the proposition is proved. [NB $\nabla \equiv$ grad]

Corollary 1   *The energy equation for a particle moving solely under the influence of a central force can be written*

$$\tfrac{1}{2}m\dot{\mathbf{r}}^2 - \int F(r)\mathrm{d}r = E. \tag{3.18}$$

Proof   Substitute for the potential energy $V(r) = -\int F(r)\mathrm{d}r$ in the energy equation (3.16).

Or

More directly, write down Newton's second law and take the scalar product with $\dot{\mathbf{r}}$

$$m\dot{\mathbf{r}}.\ddot{\mathbf{r}} = F(r)\hat{\mathbf{r}}.\dot{\mathbf{r}} = \frac{F(r)}{r}\mathbf{r}.\dot{\mathbf{r}}.$$

However

$$2\mathbf{r}.\dot{\mathbf{r}} = \frac{\mathrm{d}}{\mathrm{d}t}(\mathbf{r}.\mathbf{r}) = \frac{\mathrm{d}}{\mathrm{d}t}(r^2) = 2r\dot{r}$$

and

$$\tfrac{1}{2}m\dot{\mathbf{r}}^2 = \int \frac{F(r)}{r}r\dot{r}\,\mathrm{d}r + E$$

$$\Rightarrow \quad \tfrac{1}{2}m\dot{\mathbf{r}}^2 - \int F(r)\mathrm{d}r = E.$$

Corollary 2   *In the case that the central force acting on the particle is gravitational due to another particle the potential energy is*

$$V = -\frac{GMm}{r}$$

and the energy equation becomes

$$\frac{1}{2}m\dot{r}^2 - \frac{GMm}{r} = E. \qquad (3.19)$$

Example 1   Suppose that a rocket mass m is launched from the surface of the Earth and that its velocity relative to an inertial frame at the centre of the Earth is **V**. Ignoring the effect of the atmosphere find the condition that the rocket escapes from the Earth's pull.

Using equation (3.19) above we have

$$\frac{1}{2}m\dot{r}^2 - \frac{GMm}{r} = \frac{1}{2}mV^2 - \frac{GMm}{R}$$

where $V = |\mathbf{V}|$ and R is the radius of the Earth.

By 'the rocket escapes' we mean that the rocket escapes from the influence of the Earth altogether. Hence we require that $r \to \infty$. Since $\dot{r}^2$ is always positive this means that

$$\frac{1}{2}mV^2 - \frac{GMm}{R} \geq 0.$$

The minimum escape velocity is

$$V_e = \sqrt{\frac{2GM}{R}}$$

in any direction that does not bury the rocket in the ground! This result indicates that under realistic assumptions the most efficient place from which to launch a rocket is as close to the equator as possible. (Why?)

Example 2   Suppose that a small satellite is launched with velocity V relative to the centre of the Earth along the circular orbit of a space station which is orbiting the Earth at a distance D from its centre. If in the subsequent motion it is known that the satellite does not escape from the Earth's pull find the maximum distance of the satellite from the centre of the Earth.

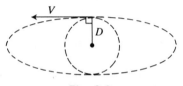

Fig. 3.3

Again using eqn (3.19) we have

$$\frac{1}{2}m\dot{r}^2 - \frac{GMm}{r} = \frac{1}{2}mV^2 - \frac{GMm}{D}$$

so that if the satellite does not escape then $V^2 < 2GM/D$.

For such a particle the angular momentum is conserved and eqn (2.12) holds

$$r^2\dot{\theta} = h = DV$$

from initial conditions. Also $\dot{\mathbf{r}}^2 = \dot{r}^2 + r^2\dot{\theta}^2$ giving

$$\frac{1}{2}m(\dot{r}^2 + r^2\dot{\theta}^2) - \frac{GMm}{r} = \frac{1}{2}mV^2 - \frac{GMm}{D}.$$

Hence

$$\dot{r}^2 + \frac{D^2V^2}{r^2} - \frac{2GM}{r} = V^2 - \frac{2GM}{D}.$$

The maximum value of r occurs when $\dot{r} = 0$. We get a quadratic equation for r one of whose roots is D. The other root is given by

$$r = \frac{D^2V^2}{2GM - DV^2}.$$

This is the greatest distance which the satellite achieves from the centre of the Earth.

3.5 Motion of a particle on a surface under gravity

The above theory applies neatly to consideration of a particle constrained to move on a surface in some way, the inertial frame being stationary on the surface of the Earth so that the distance travelled by the particle is very small, e.g. a ball-bearing sliding on the inside of a bowl *or* a spherical pendulum. If the reaction between the particle and the surface can be taken to be normal to the surface then it must be perpendicular to the velocity of the particle. This means that, if the frictional force is negligible, the energy can be shown to be conserved.

In what follows in this section we will use **q** to denote the position vector of the particle in order to avoid confusion. We will use cylindrical polar coordinates (r, θ, z) where r is the distance from the Oz axis, *not* the magnitude of the position vector.

Proposition 3.6 *If a particle is moving under gravity* **g** *on a surface in such a way that the reaction between the particle and the surface is normal to the surface then energy is conserved.*

Proof   On choosing some convenient origin O we can write, using Newton's second law,

$$m\ddot{\mathbf{q}} = m\mathbf{g} + \mathbf{R}. \tag{3.20}$$

Since $\dot{\mathbf{q}}.\mathbf{R} = 0$, taking the scalar product with $\dot{\mathbf{q}}$ and integrating gives conservation of energy

$$\tfrac{1}{2}m\dot{\mathbf{q}}^2 - m\mathbf{g}.\mathbf{q} = E. \tag{3.21}$$

In the particular case where the surface is a surface of revolution with its axis of symmetry vertical, conservation of energy together with conservation of angular momentum about the vertical axis gives a particularly neat method of tackling problems using cylindrical polar coordinates r, θ, z such that

$$x = r\cos\theta \qquad y = r\sin\theta$$

and z denotes the Cartesian coordinate as usual. (See the Appendix.)

SMOOTH SURFACE OF REVOLUTION WITH VERTICAL AXIS

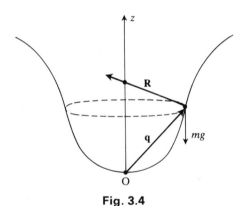

Fig. 3.4

Choose an origin O on the axis of symmetry as shown above.

Proposition 3.7   *The component of angular momentum along the axis of symmetry is constant.*

Proof   Take the vector product of the equation given by Newton's second law

$$m\mathbf{q} \wedge \ddot{\mathbf{q}} = m\mathbf{q} \wedge \mathbf{g} + \mathbf{q} \wedge \mathbf{R}.$$

Then take the scalar product with the vertical vector **k**

$$\mathbf{k}.m\mathbf{q} \wedge \ddot{\mathbf{q}} = \mathbf{k}.m\mathbf{q} \wedge \mathbf{g} + \mathbf{k}.\mathbf{q} \wedge \mathbf{R}.$$

Since **k** and **g** are parallel $\mathbf{k}.m\mathbf{q} \wedge \mathbf{g} = 0$, and by symmetry **R**, **k**, and **q** are coplanar so that $\mathbf{k}.\mathbf{q} \wedge \mathbf{R} = 0$ also. Hence

$$\mathbf{k}.m\mathbf{q} \wedge \ddot{\mathbf{q}} = 0.$$

Integrating gives

$$\mathbf{k}.\mathbf{L_O} = \mathbf{k}.m\mathbf{q} \wedge \dot{\mathbf{q}} = \text{constant.} \tag{3.22}$$

The component of angular momentum along the vertical axis is constant.

Corollary   *Using cylindrical polar coordinates*

$$mr^2\dot{\theta} = h \tag{3.23}$$

where h is a constant.

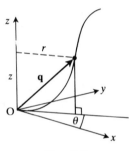

Fig. 3.5

Proof   Since $\mathbf{L_O} = \mathbf{q} \wedge m\dot{\mathbf{q}}$ and

$$\mathbf{q} = r\hat{\mathbf{r}} + z\mathbf{k}$$
$$\dot{\mathbf{q}} = \dot{r}\hat{\mathbf{r}} + r\dot{\theta}\hat{\boldsymbol{\theta}} + \dot{z}\mathbf{k},$$

the component of $\mathbf{L_O}$ along **k** is $mr^2\dot{\theta}$ and the corollary is proved.

Example   *The spherical pendulum.*

So long as the string remains taut the pendulum bob is constrained to move on the surface of a sphere, centre O radius a. We will use cylindrical polar coordinates with Oz pointing vertically downwards and r being the distance of the bob from this axis. It is clear that the tension T acts along

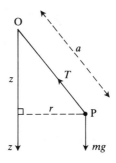

Fig. 3.6

the string, perpendicular to the velocity of the bob. Equations (3.21) and (3.23) both apply:
(3.21) gives

$$\tfrac{1}{2}m(\dot{r}^2 + r^2\dot{\theta}^2 + \dot{z}^2) - mgz = E \qquad (3.24)$$

(3.23) becomes

$$mr^2\dot{\theta} = mh, \quad \text{constant} \qquad (3.25)$$

where we have used

$$\mathbf{q} = r\hat{\mathbf{r}} + z\mathbf{k}$$

and

$$\dot{\mathbf{q}} = \dot{r}\hat{\mathbf{r}} + r\dot{\theta}\hat{\boldsymbol{\theta}} + \dot{z}\mathbf{k}.$$

Note that r and z are not independent variables since

$$r^2 + z^2 = a^2. \qquad (3.26)$$

Problem. Suppose that the pendulum bob is set in motion with a horizontal velocity V, the string also being horizontal and taut. Find the greatest depth reached by the bob below O, assuming that the string remains taut.

Solution. When projected the bob is on the same horizontal level as O and we have $z = 0$, $\dot{z} = 0$, $r\dot{\theta} = V$ when $t = 0$:
From (3.24)

$$\tfrac{1}{2}m(\dot{r}^2 + r^2\dot{\theta} + \dot{z}^2) - mgz = \tfrac{1}{2}mV^2.$$

From (3.25)

$$r^2\dot{\theta} = aV$$

and on differentiating (3.26)

$$r\dot{r} + z\dot{z} = 0.$$

We can use these equations and (3.26) to give the following equation for z, $\dot{z}$:

$$\frac{a^2\dot{z}^2}{a^2 - z^2} + \frac{a^2 V^2}{a^2 - z^2} - 2gz = V^2$$

and with a little rearrangement

$$a^2\dot{z}^2 = -2g\left(z^3 + \frac{V^2}{2g}z^2 - za^2\right)$$

or

$$a^2\dot{z}^2 = -2gz(z - z_1)(z + z_2)$$

where

$$z_1 = \sqrt{\left(\frac{V^4}{16g^2} + a^2\right)} - \frac{V^2}{4g}$$

and

$$z_2 = \sqrt{\left(\frac{V^4}{16g^2} + a^2\right)} + \frac{V^2}{4g}.$$

$\dot{z}^2$ must remain positive and so $0 < z < z_1$, showing that the bob falls initially and reaches a depth of z_1 below O, remembering that z was measured downwards.

The behaviour of the pendulum bob is shown in Fig. 3.7 below.

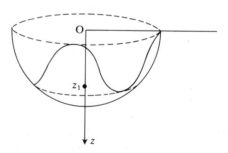

Fig. 3.7

This example concludes the investigation of conservation of energy for a single particle. We shall see that it is an important concept for systems of particles and for rigid bodies. It leads on to Lagrange's equations and a very elegant method of extending the mechanical model first put forward by Newton.

Exercises:
Chapter 3

1. A particle mass m is attached to two identical springs each of spring constant k and natural length a. One spring is attached at its other end to a fixed point A, the other spring to a fixed point B such that A is a height $4a$ vertically above B. The particle can move in the vertical line AB. If $\frac{1}{2}mg < ak$ find the position in which the particle can remain at rest and show that in this equilibirum position both springs are stretched.

    Write down the energy equation if the particle is released from rest when the lower spring is at its natural length. Show that the particle next has zero velocity at a height $3a - (mg/k)$ above B.

2. A smooth wire is in the shape of a helix, $x = a\cos\theta$, $y = a\sin\theta$, $z = b\theta$. A small bead slides on the wire whose central axis Oz is vertical. If the bead starts at a height $z = b$ from rest, write down the equation of conservation of energy and hence calculate the time taken for the bead to reach the plane $z = 0$.

3. Show that if a particle of mass m is acted on by a central force $F(r)$ directed towards a fixed point O, r being the distance from O, then the expression

$$\tfrac{1}{2}mv^2 + \int F(r)\mathrm{d}r$$

is constant throughout, where v is the velocity of the particle.

    A comet, travelling with speed V, is a very great distance from a star. If the path of the comet were unaffected by the presence of the star its closest distance of approach would be d. However, the comet experiences an attractive force γ/r^2 per unit mass directed towards the star. Calculate the actual distance of closest approach.

(Oxford)

4. A lunar orbiter is executing a circular orbit about the Moon with period 2 hours. Find the radius of the orbit.

    Derive the equation of conservation of energy for a satellite orbiting the Moon.

    A lunar module is propelled from the surface of the Moon with the intention of making a rendezvous with the orbiter. Its motors fire for a very short burst giving it an initial speed V (relative to an inertial frame at the centre of the Moon). What is the minimum speed V required to reach the circular orbit if its motors are not fired again?
    $[G = 6.67 \times 10^{-11}$ in SI units, $M_{\text{moon}} = 7.35 \times 10^{22}$ kg, $R_{\text{moon}} = 1.74 \times 10^6$m.$]$

5. A smooth surface of revolution is hyperbolic with equation $z = a^2/r$, the axis Oz pointing vertically downwards and r, θ, z being cylindrical polar co-ordinates. A small particle mass m slides on the interior of the surface. Establish the equations

$$r^2\dot\theta = h$$

$$\tfrac{1}{2}m(\dot r^2 + r^2\dot\theta^2 + \dot z^2) - mgz = E.$$

If the particle is set in motion with horizontal velocity $a\omega$ along the surface and at a depth $z = a$ below O derive an equation involving z and $\dot{z}$ only. Prove that circular motion will occur if $a\omega^2 = g$, and that otherwise the particle rises or falls according to the sign of $a\omega^2 - g$.

6. In a practical demonstration a model is designed to show deflection of a comet by the Sun, as follows. A smooth surface of revolution is constructed whose cross-sectional shape is hyperbolic, $z = a^2/r$ in cylindrical polar coordinates, the z axis pointing downwards. A small shiny ball-bearing is projected across the surface at a good distance $r = b$ from the central axis. The path of the ball-bearing is recorded in plan view from above. Assuming no friction find a first-order differential equation for $\dot{r}$ in terms of r and physical constants only. Compare your equation with a similar first-order equation for a comet moving under the influence of the Sun. Under what circumstances would you expect the two paths to display similar properties?

7. The Lennard-Jones potential

$$V(r) = 4e_0\left[\left(\frac{a}{r}\right)^{12} - \left(\frac{a}{r}\right)^{6}\right]$$

can arise when considering intermolecular forces. Use this potential in a classical central force problem, writing down the equation of conservation of energy. Sketch a graph of the potential, remembering $r > 0$. In any possible motion prove that the energy E satisfies $E \geq -e_0$.

If the angular momentum is initially zero show that there is a minimum distance of the particle from the centre of the force in any motion. Show that if the energy $E > 0$ there is no maximum distance but that if $E < 0$ there is. What can you say if the angular momentum is not zero?

8. The equation of motion of a particle of unit mass is

$$\ddot{\mathbf{r}} = -\frac{\lambda^2}{r^6}\mathbf{r}.$$

Show that if the particle starts at a point where $r = a$ with speed $\lambda/\sqrt{(2)}a^2$ in any direction then

$$\frac{1}{2}\left[\frac{h^2}{r^4}\left(\frac{dr}{d\theta}\right)^2 + \frac{h^2}{r^2}\right] - \frac{\lambda^2}{4r^4} = 0$$

where $r^2\dot{\theta} = h$. Hence show that the path of the particle is part of a curve which passes through the origin $r = 0$.

4 Rotating frames

4.1 Introduction

When considering motion close to the surface of the Earth we have as yet no formalism which takes into account the rotation of the Earth about its north–south axis. When heavy artillery fire takes place over a range of 30 km or more the effect of the Earth's rotation is detectable. In Chapter 1 we saw that the choice of fixed axes in an inertial frame S did not influence the way in which we used Newton's laws, nor, more importantly, did a change to a new frame moving with uniform velocity with respect to S. However, if the new frame is specified by axes which are rotating with respect to S we have strong evidence to suggest the frame is not inertial. Most of us have visited the fairground. Some brave souls may even have paid for a 'go' on the 'Wall' in which the joy-riding clients stand with their backs to the interior wall of a large drum. The drum spins about its central axis. At maximum speed the floor of the drum is removed leaving the incumbents pinned to the wall. If the drum were not spinning everyone would fall down when the floor was removed. The natual frame of reference as seen by the clients in the drum is a frame fixed in that drum. If this frame were known to be an inertial frame, applying Newton's second law in it would tell us that when the floor was removed the people would fall down the wall. The conclusion must be that a frame spinning with respect to an inertial frame is *not itself inertial*. Nevertheless it is often convenient to use a rotating frame of reference. The intention in the remainder of this chapter is to set up an equation derived from Newton's second law in an inertial frame so that we are able to use rotating frames of reference.

4.2 Two-dimensional rotating frame

First we shall consider the case of a basic two-dimensional rotation.

Consider a two-dimensional frame e_1, e_2 which is rotating with respect to a frame i, j as shown in Fig. 4.1 where θ, the angle between i and e_1, is time dependent. Then

$$e_1 = \cos\theta i + \sin\theta j \qquad e_2 = -\sin\theta i + \cos\theta j$$

and

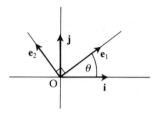

Fig. 4.1

$$\dot{\mathbf{e}}_1 = -\dot{\theta}\sin\theta\mathbf{i} + \dot{\theta}\cos\theta\mathbf{j}$$
$$\dot{\mathbf{e}}_2 = -\dot{\theta}\cos\theta\mathbf{i} - \dot{\theta}\sin\theta\mathbf{j}.$$

These last two equations can be written alternatively as

$$\dot{\mathbf{e}}_1 = \dot{\theta}\mathbf{k} \wedge \mathbf{e}_1 \qquad \dot{\mathbf{e}}_2 = \dot{\theta}\mathbf{k} \wedge \mathbf{e}_2$$

where $\mathbf{k}$ is the unit vector perpendicular to the plane as usual. It is also clear that if $\mathbf{e}_1, \mathbf{e}_2, \mathbf{e}_3$ are a right-handed set of unit vectors then $\mathbf{e}_3 = \mathbf{k}$, and

$$\dot{\mathbf{e}}_3 = \dot{\theta}\mathbf{k} \wedge \mathbf{e}_3 = \mathbf{0}$$

which gives

$$\dot{\mathbf{e}}_i = \dot{\theta}\mathbf{k} \wedge \mathbf{e}_i \quad \text{for } i = 1, 2, 3.$$

We call the vector $\dot{\theta}\mathbf{k}$ the *angular velocity* vector associated with the rotation of the moving frame with respect to $\mathbf{i}, \mathbf{j}, \mathbf{k}$. Notice that the *direction* of the angular velocity vector is perpendicular to the plane in which the rotation takes place. We can think of this direction as the *axis of rotation*.

The question now arises as to whether such an angular velocity vector always exists in the case of a general three-dimensional rotation.

4.3   Angular velocity vector

Geometrically a rotation involves turning a frame through some angle θ about an axis parallel, say, to a unit vector $\mathbf{n}$. For example, between midnight and mid-day a frame fixed in the Earth at its centre has turned through an angle of π radians about the polar axis. Intuitively we can see that the axis of rotation is the polar axis. The inertial frame (to a good approximation over half a day) is a frame with origin at the centre of the Earth fixed relative to the Sun and stars.

However, in the most general case not only can a rotating frame continuously turn about an axis of rotation relative to the inertial frame,

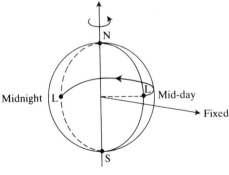

Fig. 4.2

but also the axis of rotation can change direction with time. The angular velocity, if it exists, has variable magnitude and direction.

There are many proofs of the existence of an angular velocity vector. We give two alternatives below.

Proposition 4.1

Suppose that a frame S' is rotating with respect to a frame S and that the two frames have a common origin. Then the angular velocity vector $\boldsymbol{\omega}$ exists and gives the three equations

$$\dot{\mathbf{e}}_i' = \boldsymbol{\omega} \wedge \mathbf{e}_i' \qquad (4.1)$$

where the dot indicates rate of change with respect to S and $\mathbf{e}_1'$, $\mathbf{e}_2'$, $\mathbf{e}_3'$ are unit vectors along the axes chosen in S'.

The first proof given is the standard proof. The second proof is included so that students familiar with matrices can see a connection between the orthogonal matrix U, which gives one frame of reference in terms of the other at time t, and the angular velocity. Omit proof 2 if you are unfamiliar with the material.

Proof 1

For any frame S' with unit vectors $\mathbf{e}_1', \mathbf{e}_2', \mathbf{e}_3'$ along the axes we know that

$$\mathbf{a} = \sum_1^3 (\mathbf{a}.\mathbf{e}_i')\mathbf{e}_i'.$$

Now suppose that $\mathbf{x}$ is any vector which is fixed in S'. On differentiating in the frame S' we have

$$\left(\frac{d\mathbf{x}}{dt}\right)' = 0.$$

Then for any such vector, $\mathbf{x} = \sum (\mathbf{x}.\mathbf{e}_i')\mathbf{e}_i'$, we must have $\mathbf{x}.\mathbf{e}_i' = $ constant since $\mathbf{x}$ is fixed in S':

(*) $$\Rightarrow \quad \dot{\mathbf{x}} = \sum (\mathbf{x}.\mathbf{e}_i')\dot{\mathbf{e}}_i'.$$

Also $\dot{\mathbf{x}} = \Sigma(\dot{\mathbf{x}}.\mathbf{e}'_i)\mathbf{e}'_i$ using the expression for $\mathbf{a} = \dot{\mathbf{x}}$ above. But $\mathbf{x}.\mathbf{e}'_i = $ constant $\Rightarrow \dot{\mathbf{x}}.\mathbf{e}'_i + \mathbf{x}.\dot{\mathbf{e}}'_i = 0$

(∗∗)
$$\Rightarrow \quad \dot{\mathbf{x}} = -\sum(\mathbf{x}.\dot{\mathbf{e}}'_i)\mathbf{e}'_i.$$

By adding (∗) and (∗∗) we can write

$$\dot{\mathbf{x}} = \tfrac{1}{2}\sum[(\mathbf{x}.\mathbf{e}'_i)\dot{\mathbf{e}}'_i - (\mathbf{x}.\dot{\mathbf{e}}'_i)\mathbf{e}'_i]$$

$$\dot{\mathbf{x}} = \tfrac{1}{2}\sum(\mathbf{e}'_i \wedge \dot{\mathbf{e}}'_i) \wedge \mathbf{x}.$$

We define

$$\boldsymbol{\omega} = \tfrac{1}{2}\sum(\mathbf{e}'_i \wedge \dot{\mathbf{e}}'_i)$$

$$\Rightarrow \quad \dot{\mathbf{x}} = \boldsymbol{\omega} \wedge \mathbf{x} \quad \forall \text{ vectors } \mathbf{x}.$$

In particular this also holds for $\mathbf{e}'_i$ and the proof is concluded.

Proof 2   Suppose that $\mathbf{e}_i$ are unit vectors along axes fixed in the frame S as shown in Fig. 4.3. At any fixed time $t\{\mathbf{e}'_i\}$ are a fixed set of orthogonal unit vectors along axes in S′. Hence ∃ an orthogonal matrix $U(t)$ such that

$$\mathbf{e}'_i = U(t)\mathbf{e}_i \quad \text{and} \quad U(t)U(t)^{\mathrm{T}} = I$$

where in S we have

$$\mathbf{e}_1 = \begin{bmatrix} 1 \\ 0 \\ 0 \end{bmatrix}$$

and so on.

Then $U(t) = [U_{ij}]$, where $U_{ij} = \mathbf{e}_i.\mathbf{e}'_j$, giving the components of $\mathbf{e}'_i$ with respect to $\{\mathbf{e}_i\}$:

$$\mathbf{e}'_i = U(t)\mathbf{e}_i$$

$$\Rightarrow \quad \dot{\mathbf{e}}'_i = \dot{U}\mathbf{e}_i, \text{ since } \dot{\mathbf{e}}_i = 0 \text{ in S}$$

$$\Rightarrow \quad \dot{\mathbf{e}}'_i = \dot{U}U^{\mathrm{T}}\mathbf{e}'_i.$$

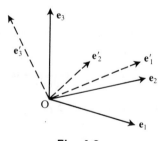

Fig. 4.3

But $UU^{\mathrm{T}} = I$ gives $\dot{U}U^{\mathrm{T}} + U\dot{U}^{\mathrm{T}} = 0$

$$\Rightarrow \quad (\dot{U}U^{\mathrm{T}})^{\mathrm{T}} = -(\dot{U}U^{\mathrm{T}}).$$

This last equation states that the matrix $\dot{U}U^{\mathrm{T}}$ is antisymmetric. Hence we can write

$$\dot{U}U^{\mathrm{T}} = \begin{bmatrix} 0 & -\omega_3 & \omega_2 \\ \omega_3 & 0 & -\omega_1 \\ -\omega_2 & \omega_1 & 0 \end{bmatrix}$$

choosing suitable labels for the components of $\dot{U}U^{\mathrm{T}}$.

Note that for any vector $\mathbf{a}$, with this choice of labels for the entries of $\dot{U}U^{\mathrm{T}}$,

$$\dot{U}U^{\mathrm{T}}\mathbf{a} = \boldsymbol{\omega} \wedge \mathbf{a}.$$

Writing $\mathbf{a}$ as a column vector with coordinates relative to $\mathbf{e}_1, \mathbf{e}_2, \mathbf{e}_3$:

$$\Rightarrow \quad \dot{U}U^{\mathrm{T}}\mathbf{e}_i' = \boldsymbol{\omega} \wedge \mathbf{e}_i'$$

$$\Rightarrow \quad \dot{\mathbf{e}}_i' = \boldsymbol{\omega} \wedge \mathbf{e}_i'.$$

Corollary 1   *The angular velocity vector is unique.*

Proof   Suppose not. Then $\dot{\mathbf{x}} = \boldsymbol{\omega} \wedge \mathbf{x}$ and $\dot{\mathbf{x}} = \boldsymbol{\omega}' \wedge \mathbf{x} \quad \forall \mathbf{x}$

$$\Rightarrow \quad \mathbf{0} = (\boldsymbol{\omega} - \boldsymbol{\omega}') \wedge \mathbf{x} \quad \forall \mathbf{x}$$

$$\Rightarrow \quad \mathbf{0} = (\boldsymbol{\omega} - \boldsymbol{\omega}').$$

Corollary 2   *Rotating Axes Theorem*

$$\frac{\mathrm{d}\mathbf{x}}{\mathrm{d}t} = \left(\frac{\mathrm{d}\mathbf{x}}{\mathrm{d}t}\right)' + \boldsymbol{\omega} \wedge \mathbf{x} \tag{4.2}$$

where $\boldsymbol{\omega}$ is the angular velocity of S' with respect to S.

Proof   In terms of the rotating axes for any vector $\mathbf{x}$ we have

$$\mathbf{x} = \sum x_i \mathbf{e}_i'$$

$$\Rightarrow \quad \dot{\mathbf{x}} = \sum \dot{x}_i \mathbf{e}_i' + \sum x_i \dot{\mathbf{e}}_i'$$

$$\Rightarrow \quad \dot{\mathbf{x}} = \sum \dot{x}_i \mathbf{e}_i' + \sum x_i \boldsymbol{\omega} \wedge \mathbf{e}_i'.$$

The first term on the right is the rate of change with respect to the rotating frame and by using the distributive property of the vector product we see that the second term is precisely $\boldsymbol{\omega} \wedge \mathbf{x}$. This proves the theorem.

The rotating axes theorem leads us to the formulation of a law of motion in a frame rotating with respect to an inertial frame and as such is the most important result in this section.

One further result will help us to find the angular velocity in specific instances.

Proposition *If S" rotates with angular velocity ω' with respect to S' and S' rotates with*
4.2 *angular velocity ω with respect to S, then S" rotates with angular velocity*
$\omega + \omega'$ with respect to S.

Proof

$$\left(\frac{d\mathbf{x}}{dt}\right)' = \left(\frac{d\mathbf{x}}{dt}\right)'' + \omega' \wedge \mathbf{x}$$

$$\left(\frac{d\mathbf{x}}{dt}\right) = \left(\frac{d\mathbf{x}}{dt}\right)' + \omega \wedge \mathbf{x}$$

$$= \left(\frac{d\mathbf{x}}{dt}\right)'' + \omega' \wedge \mathbf{x} + \omega \wedge \mathbf{x}.$$

Hence result.

One useful point to note is that the direction of the angular velocity can be found from a right-hand rule. If you curl the fingers on your right hand along the directions in which the points of the rotating frame are moving, then your thumb points in the direction of the angular velocity, considering the hand in the position required for thumbing a lift.

Examples   (a) A cylinder rolls without slipping along a fixed plane, its centre of mass G moving with constant speed v. Find the angular velocity of axes fixed in the cylinder at the centre of mass. If G moves a distance δx then the point of contact moves δx. Since there is no slipping the arc $P_1 P_2$ on the rim is the same length as $P_1 P_2$ on the plane:

$$\Rightarrow \quad \delta x = a\delta\theta$$

$$\Rightarrow \quad v = a\dot{\theta}.$$

The axis of rotation is $\mathbf{e}$, a unit vector along the axis of symmetry into the paper, and so we have

$$\omega = \dot{\theta}\mathbf{e} = \frac{v}{a}\mathbf{e}.$$

Notice that in this case the origin of the rotating axes is in motion but we can overcome the problem by considering a parallel set of unit vectors at the origin of an inertial frame as shown in Fig. 4.4.

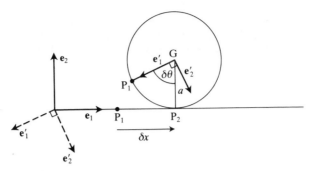

Fig. 4.4

(b) When using spherical polar coordinates r, θ, ϕ, we have three unit vectors which form a right-handed triad consisting of three vectors of unit length, mutually perpendicular. These vectors are denoted by $\hat{\mathbf{r}}$, $\hat{\boldsymbol{\theta}}$, $\hat{\boldsymbol{\phi}}$ as shown in Fig. 4.5 below (see the Appendix). Consider a parallel set at the origin O.

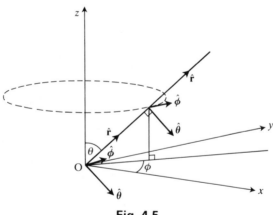

Fig. 4.5

As θ and ϕ change the vectors rotate. We can find the angular velocity of the frame $\hat{\mathbf{r}}$, $\hat{\boldsymbol{\theta}}$, $\hat{\boldsymbol{\phi}}$ with respect to $\mathbf{i}$, $\mathbf{j}$, $\mathbf{k}$. We do this by first considering the rotation as ϕ changes and θ remains fixed. Every point P fixed in the rotating frame moves round a circle perpendicular to Oz with angular speed $\dot{\phi}$. The axis of that rotation is Oz.

Next consider the rotation produced as θ increases and ϕ remains fixed. This time the axis of rotation is $\hat{\boldsymbol{\phi}}$ because it is the vector perpendicular to the circle on which P moves as θ alters. In fact if you consider both of these rotations in terms of the globe, changing ϕ produces a circle of latitude, changing θ produces a circle of longitude as the point P moves. Hence

$$\boldsymbol{\omega} = \dot{\theta}\hat{\boldsymbol{\phi}} + \dot{\phi}\mathbf{k}$$

where we have used the fact that angular velocities are additive (proposition 4.2).

4.4   Particle moving in a rotating frame of reference

Now we come to the crucial point–consideration of motion in a rotating frame. Suppose that a particle P is moving in a frame S' which rotates with angular velocity $\boldsymbol{\omega}$ with respect to an inertial frame S. The position vector of the particle is $\mathbf{r}$ measured from the common origin of the two frames.

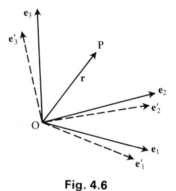

Fig. 4.6

Proposition 4.3   *If the prime denotes with respect to S' then the velocity and acceleration of the particle in S are given respectively by*

$$\frac{d\mathbf{r}}{dt} = \left(\frac{d\mathbf{r}}{dt}\right)' + \boldsymbol{\omega} \wedge \mathbf{r} \tag{4.3}$$

$$\frac{d^2\mathbf{r}}{dt^2} = \left(\frac{d^2\mathbf{r}}{dt^2}\right)' + 2\boldsymbol{\omega} \wedge \left(\frac{d\mathbf{r}}{dt}\right)' + \frac{d\boldsymbol{\omega}}{dt} \wedge \mathbf{r} + \boldsymbol{\omega} \wedge (\boldsymbol{\omega} \wedge \mathbf{r}). \tag{4.4}$$

Proof   Corollary 2 in the previous section immediately gives (4.3). In fact we can think of the differential operators in S and S' as being related by

$$\frac{d}{dt} = \left(\frac{d}{dt}\right)' + \boldsymbol{\omega} \wedge \quad .$$

To get (4.4) we differentiate as follows

$$\frac{d^2\mathbf{r}}{dt^2} = \frac{d}{dt}\left(\frac{d\mathbf{r}}{dt}\right)' + \frac{d}{dt}(\boldsymbol{\omega} \wedge \mathbf{r})$$

$$\frac{d^2\mathbf{r}}{dt^2} = \left(\frac{d^2\mathbf{r}}{dt^2}\right)' + \boldsymbol{\omega} \wedge \left(\frac{d\mathbf{r}}{dt}\right)' + \frac{d\boldsymbol{\omega}}{dt} \wedge \mathbf{r} + \boldsymbol{\omega} \wedge \frac{d\mathbf{r}}{dt}.$$

Using (4.3) to replace $d\mathbf{r}/dt$ in the last term we immediately get (4.4). Note that

$$\frac{d\boldsymbol{\omega}}{dt} = \left(\frac{d\boldsymbol{\omega}}{dt}\right)' \tag{4.5}$$

so that we may find the rate of change of the angular velocity itself with respect to either frame.

If we consider a particle moving in a frame S' rotating with angular velocity $\boldsymbol{\omega}$ with respect to an inertial frame S and acted on by a force $\mathbf{F}$, on using Newton's second law in the inertial frame and switching to the rotating frame we have

(A)
$$m\frac{d^2\mathbf{r}}{dt^2} = m\left[\left(\frac{d^2\mathbf{r}}{dt^2}\right)' + 2\boldsymbol{\omega} \wedge \left(\frac{d\mathbf{r}}{dt}\right)' + \frac{d\boldsymbol{\omega}}{dt} \wedge \mathbf{r} + \boldsymbol{\omega} \wedge (\boldsymbol{\omega} \wedge \mathbf{r})\right] = \mathbf{F}.$$

If in addition $\dot{\boldsymbol{\omega}} = \mathbf{0}$ then

$$m\left(\frac{d^2\mathbf{r}}{dt^2}\right)' = -2m\boldsymbol{\omega} \wedge \left(\frac{d\mathbf{r}}{dt}\right)' - m\boldsymbol{\omega} \wedge (\boldsymbol{\omega} \wedge \mathbf{r}) + \mathbf{F}.$$

The first term on the right-hand side used to be called the 'Coriolis force' whereas the second term was referred to as the 'centrifugal force'. We can now see that they arise naturally as part of the acceleration with respect to the inertial frame expressed in terms of the acceleration and velocity in the rotating frame. In the examples below we shall use Newton's law in the form (A) above rather than introduce fictitious forces.

Example 1   A bead slides on a smooth helix whose central axis is vertical. The helix is forced to rotate about its central axis with constant angular velocity ω. Find the equation of motion of the bead relative to the helix.

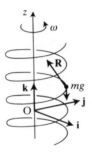

Fig. 4.7

A helix (shaped like a spring) can be parametrized as follows:

$$\mathbf{r} = a\cos\theta\mathbf{i} + a\sin\theta\mathbf{j} + b\theta\mathbf{k}.$$

With respect to axes $\mathbf{i}, \mathbf{j}, \mathbf{k}$ fixed in the helix we must have

$$\dot{\mathbf{r}}' = \dot{\theta}(-a\sin\theta\mathbf{i} + a\cos\theta\mathbf{j} + b\mathbf{k})$$

$$\Rightarrow \quad \ddot{\mathbf{r}}' = \ddot{\theta}(-a\sin\theta\mathbf{i} + a\cos\theta\mathbf{j} + b\mathbf{k}) - \dot{\theta}^2(a\cos\theta\mathbf{i} + a\sin\theta\mathbf{j}).$$

Newton's second law gives

$$m\ddot{\mathbf{r}} = -mg\mathbf{k} + \mathbf{R}$$

$$\Rightarrow \quad m(\ddot{\mathbf{r}}' + 2\omega\mathbf{k} \wedge \dot{\mathbf{r}}' + \omega\mathbf{k} \wedge (\omega\mathbf{k} \wedge \mathbf{r})) = -mg\mathbf{k} + \mathbf{R}. \qquad (4.6)$$

But $\dot{\mathbf{r}}'$ is tangent to the helix and hence perpendicular to $\mathbf{R}$, which is normal to the helix. Looking at the expression for $\dot{\mathbf{r}}'$ we can see that

$$\mathbf{c} = (-a\sin\theta\mathbf{i} + a\cos\theta\mathbf{j} + b\mathbf{k}) = \dot{\theta}^{-1}\dot{\mathbf{r}}'$$

is also parallel to the tangent to the helix. If we take the scalar product of eqn (4.6) with this vector then this gives a scalar equation and removes $\mathbf{R}$ since $\mathbf{c}.\mathbf{R} = 0$. The second term on the left is also perpendicular to $\mathbf{c}$ which leaves two terms on the left and one on the right to evaluate. With a little manipulation

$$\ddot{\theta}(a^2 + b^2) = -gb$$

and so $\ddot{\theta}$ is constant.

Example 2   A fine smooth wire is in the shape of an ellipse in a horizontal plane. The wire rotates about a vertical axis through the centre of the ellipse, with constant angular speed ω. Show that a small bead sliding on the wire can remain at rest relative to the wire at either end of both the major and minor axes. Which of these positions is stable?

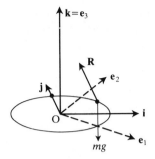

Fig. 4.8

The ellipse can be parametrized as follows:

$$\mathbf{r} = a\cos\theta\mathbf{i} + b\sin\theta\mathbf{j}$$

assuming $a > b$. Then with respect to the axes shown fixed in the wire

$$\dot{\mathbf{r}}' = \dot{\theta}(-a\sin\theta\mathbf{i} + b\cos\theta\mathbf{j}) = \dot{\theta}\frac{d\mathbf{r}}{d\theta}$$

$$\ddot{\mathbf{r}}' = \ddot{\theta}(-a\sin\theta\mathbf{i} + b\cos\theta\mathbf{j}) + \dot{\theta}^2(-a\cos\theta\mathbf{i} - b\sin\theta\mathbf{j}).$$

Using Newton's second law we have

$$m\frac{d^2\mathbf{r}}{dt^2} = -mg\mathbf{k} + \mathbf{R}$$

and so

$$m\ddot{\mathbf{r}} = m[\ddot{\mathbf{r}}' + 2\omega\mathbf{k} \wedge \dot{\mathbf{r}}' + \omega\mathbf{k} \wedge (\omega\mathbf{k} \wedge \mathbf{r})] = -mg\mathbf{k} + \mathbf{R}.$$

We know that $d\mathbf{r}/d\theta$ is tangential to the ellipse and is perpendicular to $\mathbf{R}$ and $\mathbf{k}$. Taking the scalar product with this tangent vector gives, after manipulation,

$$\ddot{\theta}(a^2\sin^2\theta + b^2\cos^2\theta) + \dot{\theta}^2(a^2 - b^2)\sin\theta\cos\theta$$
$$+ \omega^2(a^2 - b^2)\sin\theta\cos\theta = 0.$$

If the bead is at rest relative to the wire then $\ddot{\theta} = \dot{\theta} = 0$ giving

$$\theta = 0, \pi \quad \text{or} \quad \tfrac{1}{2}\pi, \tfrac{3}{2}\pi.$$

The bead can stay put at any one of the four ends of the axes.

Close to $\theta = 0$ set $\theta = \varepsilon$, a small non-constant parameter. Assuming $\dot{\varepsilon}$ is small also the equation becomes

$$b^2\ddot{\varepsilon} + \omega^2(a^2 - b^2)\varepsilon = O(\varepsilon^2)$$

that is, simple harmonic motion since $a > b$. This is a stable equilibrium position relative to the wire.

Close to $\theta = \frac{1}{2}\pi$ set $\theta = \frac{1}{2}\pi + \varepsilon$. Then $\sin(\frac{1}{2}\pi + \varepsilon) = 1$, $\cos(\frac{1}{2}\pi + \varepsilon) = -\varepsilon$ approximately. We have

$$a^2\ddot{\varepsilon} - \omega^2(a^2 - b^2)\varepsilon = O(\varepsilon^2)$$

and hence since the solutions are exponential the equilibrium position (relative to the wire) is unstable.

Example 3   *The 'Wall'.* Now consider the 'Wall' at the fairground. It consists of a very large drum which can revolve about its central vertical axis. We will model a person, standing against the interior wall of the drum, as a particle. Suppose that the position vector of the particle from the central axis is $a\mathbf{e}$ where a is the radius of the drum and $\mathbf{e}$ is a unit vector fixed in the frame of the drum. Assume that the drum rotates with constant angular velocity $\omega\mathbf{k}$ about its central axis. The forces on the particle are the force due to gravity, a normal reaction $\mathbf{R}$, and a frictional force $\mathbf{F}$ parallel to the wall.

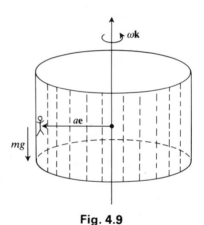

Fig. 4.9

In the rotating frame the position of the particle is constant. Hence we have

$$\mathbf{r} = a\mathbf{e} \qquad \left(\frac{d\mathbf{r}}{dt}\right)' = 0 \qquad \left(\frac{d^2\mathbf{r}}{dt^2}\right)' = 0$$

giving

$$m\frac{d^2\mathbf{r}}{dt^2} = m\omega\mathbf{k} \wedge (\omega\mathbf{k} \wedge a\mathbf{e}) = -mg\mathbf{k} + \mathbf{R} + \mathbf{F}. \qquad (4.7)$$

The only force in the direction of e is **R**, the normal reaction. Evaluating the triple vector product we have

$$-ma\omega^2\mathbf{e} = \mathbf{R} \quad and \quad 0 = -mg\mathbf{k} + \mathbf{F}. \tag{4.8}$$

The magnitude of the normal reaction is thus $ma\omega^2$. Notice that the left-hand side of the first equation is simply the *acceleration towards the centre*, of magnitude $a\omega^2$. A simple model of friction suggests that the maximum friction which can be developed between two objects is of the form $\mu|\mathbf{R}|$, so that in this case slipping will not occur if

$$|\mathbf{F}| \leq \mu|\mathbf{R}|.$$

Using eqns (4.8) we have

$$|\mathbf{F}| = mg \leq \mu ma\omega^2. \tag{4.9}$$

In words: *sufficient friction can be developed to prevent the person sliding down the wall provided the drum is revolving fast enough.*

4.5   Motion on the surface of the Earth

For a projectile such as a short-range missile launched from the surface of the Earth the natural frame of reference to use is a frame of reference at the position of launching, with one axis vertical and the other two horizontal pointing due east and north. Consider a frame of reference S situated at the centre of the Earth which does not rotate with the Earth but which maintains its position relative to the Sun. Over a few hours the frame of reference S is (approximately) moving with uniform velocity with respect to the Sun and as such can be considered to be an inertial frame. The frame at the surface displays two features which ensure that it is not inertial. One property is that it rotates with the Earth. The other property is that its origin O' moves on a circle about the NS polar axis and hence is accelerating. Both these must be taken into account. Suppose that **i, j, k** are axes as shown in Fig. 4.10, **i** into the page, rotating in the frame origin O' situated at the surface. Then if **r** is the position of the projectile measured from O' the position of the particle from O is $\mathbf{R} = \mathbf{a} + \mathbf{r}$ where $\mathbf{a} = \mathrm{OO}'$. The equation of motion of the projectile is

$$m\frac{d^2\mathbf{R}}{dt^2} = m\left[\left(\frac{d^2\mathbf{R}}{dt^2}\right)' + 2\omega\mathbf{n} \wedge \left(\frac{d\mathbf{R}}{dt}\right)' + \omega\mathbf{n} \wedge (\omega\mathbf{n} \wedge \mathbf{R})\right]$$

$$= m\mathbf{g} \tag{4.10}$$

where the prime indicates rate of change with respect to a set of axes parallel to **i, j, k** situated at origin O. These are rotating with angular

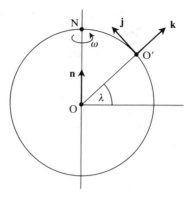

Fig. 4.10

velocity $\omega\mathbf{n}$, the angular velocity of the Earth, but now the difficulty of having an accelerating origin O' has been removed. O is also the origin of an inertial frame relative to the Sun and stars. In the rotating frame situated at O, $\mathbf{a}$ is a constant vector and (4.10) becomes

$$m\left[\left(\frac{d^2\mathbf{r}}{dt^2}\right)' + 2\omega\mathbf{n}\wedge\left(\frac{d\mathbf{r}}{dt}\right)' + \omega\mathbf{n}\wedge(\omega\mathbf{n}\wedge(\mathbf{a}+\mathbf{r}))\right] = m\mathbf{g}. \quad (4.11)$$

The term $\omega\mathbf{n}\wedge(\omega\mathbf{n}\wedge\mathbf{a})$ is in fact precisely the acceleration of O'. To a first approximation in ω which is of course very small, at 2π radians per day, we can ignore the term involving $\omega^2\mathbf{r}$ in eqn (4.11). However, the radius of the Earth $|\mathbf{a}|$ is very large and so the contribution due to the acceleration of O' cannot be discounted. We have

$$\left(\frac{d^2\mathbf{r}}{dt^2}\right)' + 2\omega\mathbf{n}\wedge\left(\frac{d\mathbf{r}}{dt}\right)' = \mathbf{g} - \omega^2\mathbf{n}\wedge(\mathbf{n}\wedge\mathbf{a}). \quad (4.12)$$

If the projectile is fired over a short range, say 25 km, then $\mathbf{g}$ can be regarded as a constant. If the range is a good deal longer then the direction of $\mathbf{g}$ will vary as $\mathbf{g}$ is directed towards O from all points on the surface. We will assume that the range is short enough so that $\mathbf{g}$ is constant. Then the expression on the right of (4.12) is called the *apparent gravity* $\mathbf{g}'$. If λ is the latitude of O' then

$$\mathbf{g}' = \mathbf{g} - \omega^2\mathbf{n}\wedge(\mathbf{n}\wedge\mathbf{a})$$

$$= -g\mathbf{k} - \omega^2(\mathbf{a}.\mathbf{n})\mathbf{n} + \omega^2\mathbf{a}$$

$$= -g\mathbf{k} - \omega^2[a\sin\lambda(\cos\lambda\mathbf{j} + \sin\lambda\mathbf{k}) - a\mathbf{k}].$$

We have

$$\mathbf{g}' = -(g - a\omega^2\cos^2\lambda)\mathbf{k} - a\omega^2\sin\lambda\cos\lambda\mathbf{j}. \quad (4.13)$$

This equation shows also that the direction of $\mathbf{g}'$ is different. The *apparent latitude* is defined to be the angle between the line of action of $\mathbf{g}'$ and the equatorial plane, giving $g' \sin \lambda' = -\mathbf{g}'.\mathbf{n} = -\mathbf{g}'.(\cos \lambda \mathbf{j} + \sin \lambda \mathbf{k})$ so that

$$g' \sin \lambda' = g \sin \lambda \quad \text{where } g' = g - a\omega^2 \cos \lambda$$

correct to first order in $a\omega^2$. Notice that the minimum value of the apparent gravity occurs at the equator and the maximum at the poles, as one might expect.

We can rewrite (4.12) as

$$\left(\frac{d^2\mathbf{r}}{dt^2}\right)' + 2\omega\mathbf{n} \wedge \left(\frac{d\mathbf{r}}{dt}\right)' = \mathbf{g}'. \tag{4.14}$$

This is the equation we need to solve in order to find the trajectory of the projectile, using the fact that ω is small.

Integrating (4.14) in the primed frame once gives

$$\left(\frac{d\mathbf{r}}{dt}\right)' + 2\omega\mathbf{n} \wedge \mathbf{r} = \mathbf{g}'t + \mathbf{C}$$

where $\mathbf{C}$ is a constant.

If we suppose that the initial velocity of the projectile is $\mathbf{V}$ relative to the Earth' surface from $\mathbf{r} = \mathbf{0}$ at $t = 0$ then $\mathbf{C} = \mathbf{V}$. Replacing $(d\mathbf{r}/dt)'$ in (4.14) by the expression found from the above equation we get

$$\left(\frac{d^2\mathbf{r}}{dt^2}\right)' + 2\omega\mathbf{n} \wedge (\mathbf{g}'t + \mathbf{V} - 2\omega\mathbf{n} \wedge \mathbf{r}) = \mathbf{g}'. \tag{4.15}$$

However, ω^2 is small and we can discard a term in (4.15) giving

$$\left(\frac{d^2\mathbf{r}}{dt^2}\right)' = \mathbf{g}' - 2\omega t\mathbf{n} \wedge \mathbf{g}' - 2\omega\mathbf{n} \wedge \mathbf{V}.$$

Integrating

$$\left(\frac{d\mathbf{r}}{dt}\right)' = \mathbf{g}'t - \omega t^2\mathbf{n} \wedge \mathbf{g}' - 2\omega t\mathbf{n} \wedge \mathbf{V} + \mathbf{D}.$$

At $t = 0$ the velocity is $\mathbf{V}$ giving $\mathbf{D} = \mathbf{V}$. Integrating again

$$\mathbf{r} = \tfrac{1}{2}\mathbf{g}'t^2 - \tfrac{1}{3}\omega t^3\mathbf{n} \wedge \mathbf{g}' - \omega t^2\mathbf{n} \wedge \mathbf{V} + \mathbf{V}t \tag{4.16}$$

where we have already used the initial conditions to give a zero constant of integration. Notice that if we put $\omega = 0$ the trajectory of the projectile reduces to the trajectory found in Chapter 1.

Exercises: Chapter 4

1. Consider a set of axes with origin O which are rotating with respect to an inertial frame with the same origin O. Two of these axes, with corresponding

directions $\mathbf{i}$, $\mathbf{j}$, are rotating within a plane which contains O so that $\mathbf{i}$ makes a variable angle θ with a fixed direction $\mathbf{e}$ in the plane. The vector $\mathbf{k}$ is the unit normal to the plane. In addition the plane rotates with angular speed Ω about a line through O parallel to $\mathbf{e}$ in the inertial frame. Write down the angular velocity of the rotating frame with respect to the inertial frame. How can you justify your answer?

2. A small stone is thrown inside a room which rotates about a central vertical axis Oz with constant angular velocity ω. The axes Ox, Oy are fixed along the floor of the room. Derive the equations of motion of the stone relative to the rotating axes showing that

$$\ddot{x} - 2\omega\dot{y} - \omega^2 x = 0$$

$$\ddot{y} + 2\omega\dot{x} - \omega^2 y = 0$$

$$\ddot{z} = -g.$$

Deduce that

$$\ddot{\zeta} + 2i\omega\dot{\zeta} - \omega^2\zeta = 0$$

where $\zeta = x + iy$. (This ignores air resistance.)

   Hence find the path of the stone in the form $\mathbf{r} = \mathbf{r}(t)$ relative to $Oxyz$, if it is thrown from a point $(a, 0, h)$ with velocity $(V\cos\alpha, 0, V\sin\alpha)$ relative to the rotating room. Compare with the path relative to the surface of the Earth. Is your answer what you would expect to find in the inertial frame?

3. A bead slides on a smooth circular hoop which is forced to rotate with constant angular velocity ω about a vertical diameter. Use a set of axes fixed in the hoop and take θ to be the angle between the downward vertical and the radius to the bead. Prove that if $g < a\omega^2$ there are four positions in which the bead can remain at rest relative to the hoop, a being the radius of the hoop. If $g > a\omega^2$ prove that the bead can perform small oscillations about the lowest point of the hoop and find the period of the oscillations.

4. Two frames of references S and S' have a common origin O and S' rotates with constant angular velocity $\boldsymbol{\omega}$ with respect to S. Prove that

$$\frac{d^2\mathbf{r}}{dt^2} = \left(\frac{d^2\mathbf{r}}{dt^2}\right)' + 2\boldsymbol{\omega} \wedge \left(\frac{d\mathbf{r}}{dt}\right)' + \boldsymbol{\omega} \wedge (\boldsymbol{\omega} \wedge \mathbf{r})$$

where $\mathbf{r}$ is the position vector of a point P measured from the origin.

   A square hoop ABCD is made of fine smooth wire and has side length $2a$. The hoop is horizontal and rotating with constant angular speed ω about a vertical axis through A. A small bead which can slide on the wire is initially at rest at the midpoint of the side BC. Choose axes fixed relative to the hoop and let y be the distance of the bead from the vertex B on the side BC. Write down the position vector of the bead in your rotating frame.

   Show that

$$\ddot{y} - \omega^2 y = 0$$

using the expression for the acceleration above. Hence find the time which the bead takes to reach a corner of the hoop.

(Oxford)

5. Write down the equations of motion of a particle near the Earth taking into account the Earth's angular velocity of rotation ω. Hence find the position vector of the particle at time t assuming general initial conditions. Use the apparent gravity $\mathbf{g}'$ in your solution.

   At a latitude λ' a shell is fired from a gun which has muzzle velocity V. Show that when the gun fires at an elevation of 45° its range when pointing due east is greater than its range when pointing due west by a distance of

   $$\frac{4\sqrt{(2)}\,\omega V^3 \cos \lambda'}{3(g')^2}$$

   assuming that terms involving ω^2 may be ignored. The curvature of the Earth's surface has also been ignored.

   (Oxford)

6. Foucault's pendulum consists of a long light inextensible string of length a and a heavy bob of mass m. The pendulum is attached to a point fixed relative to the surface of the Earth and can perform small oscillations about its equilibrium position. Let λ be the apparent latitude and $\mathbf{g}$ the apparent gravity. Suppose that $\mathbf{g} = -g\mathbf{k}$ and that $\mathbf{i}$ points due east. Take the origin O to be the equilibrium position of the bob with $Oxyz$ parallel to $\mathbf{i}$, $\mathbf{j}$, $\mathbf{k}$. If the position vector of the bob is $\mathbf{r} = x\mathbf{i} + y\mathbf{j} + z\mathbf{k}$ show that

   $$a^2 = |\mathbf{r} - a\mathbf{k}|^2$$

   and hence that z is small compared with x, y. What are the forces on the bob? Assuming that $z \approx 0$ show that if $\mathbf{s} = x\mathbf{i} + y\mathbf{j}$ then

   $$\ddot{\mathbf{s}} + 2\omega \sin \lambda \mathbf{k} \wedge \dot{\mathbf{s}} + \frac{g}{a}\mathbf{s} = \mathbf{0}.$$

7. Consider a satellite orbiting the Earth, the only force involved being the inverse square law. Prove (using results from Chapter 2) that if the satellite has a circular orbit of radius a and angular velocity $\omega\mathbf{n}$ then

   $$\omega^2 = \frac{GM}{a^3}$$

   where M is the mass of the Earth and $\mathbf{n}$ is the unit vector perpendicular to the plane containing the orbit.

   Writing the position vector of the satellite from the centre of the Earth as

   $$\mathbf{r} = \mathbf{\rho} + \mathbf{\varepsilon}$$

   where $\mathbf{\rho}$ is the position vector for the circular orbit and $\mathbf{\varepsilon}$ is a small deviation from it, show that correct to first order in $|\mathbf{\varepsilon}|$

   $$\ddot{\mathbf{\varepsilon}} = -\frac{GM}{a^3}\mathbf{\varepsilon} + \frac{3GM}{a^5}(\mathbf{\varepsilon}.\mathbf{\rho})\mathbf{\rho}.$$

   Now suppose that we use a set of axes at the centre of the Earth rotating with angualr velocity $\omega\mathbf{n}$, the same angular velocity as that of the satellite in its circular orbit. We can take $\mathbf{i}$, $\mathbf{j}$, $\mathbf{n}$ as a right-handed set of perpendicular unit vectors such that $\mathbf{\rho}$, $\mathbf{i}$ have the same direction, $\mathbf{\rho} = a\mathbf{i}$. Show that using the rotating frame the above equation becomes

$$\ddot{\boldsymbol{\varepsilon}}' + 2\omega\mathbf{n} \wedge \dot{\boldsymbol{\varepsilon}}' = -\omega^2(\mathbf{n.\varepsilon})\mathbf{n} + \frac{3GM}{a^3}(\boldsymbol{\varepsilon}.\mathbf{i})\mathbf{i}$$

where the prime denotes the rotating frame.

By setting $\boldsymbol{\varepsilon} = x\mathbf{i} + y\mathbf{j} + z\mathbf{n}$ find the general solution of this linearized vector equation and show that a disturbance solely in the direction $\mathbf{n}$ is stable to this degree of approximation. What conclusion can you draw about stability in general?

5 Many-particle systems

5.1 Motion of centre of mass

There are many binary systems of stars studied by astrophysicists. The binary system is particularly stable but true tertiary star systems are relatively rare. In fact tertiary systems often comprise a binary system together with a single star. There is no complete analytical solution to the equations satisfied by a tertiary system of star. For systems of more than two stars the equations have to be solved by numerical methods and stable solutions appear to be rare. The planetary system is of course stable but then the Sun is considerably more massive than any of the planets. We shall study the equations associated with many-particle systems very briefly.

We will look at a system of many particles in which any given pair of particles interact by exerting forces on each other. This is the first situation covered by Newton's model of the physical world in which we need to use the third law.

Newton's third law
Action and reaction are equal and opposite. (NLIII)

This tells us that if particle 1 exerts a force $\mathbf{F}$ on particle 2 then the force exerted by particle 2 on particle 1 is $-\mathbf{F}$.

What turns out to be a crucial component in solving many-particle problems is the motion of the centre of mass. Once that motion is known then attention can be paid to the motion of the system around it.

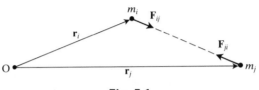

Fig. 5.1

We choose an origin in the inertial frame in which the particles move. Suppose that the ith particle has position vector $\mathbf{r}_i$ and that the force exerted by the jth particle on the ith particle is $\mathbf{F}_{ij}$. Then Newton's third law states that action and reaction are equal and opposite, so that

$$\mathbf{F}_{ij} + \mathbf{F}_{ji} = \mathbf{0}. \tag{5.1}$$

For the ith particle Newton's second law gives

$$m_i \ddot{\mathbf{r}}_i = \sum_{j \neq i} \mathbf{F}_{ij} + \mathbf{F}_i \tag{5.2}$$

where $\mathbf{F}_i$ is the *external* force on the particle, i.e. a force with arises *outside the system of particles*. In order to consider the centre of mass we need a straightforward vector definition which is easily usable.

Definition 5.1   If $M = \Sigma_i\, m_i$ is the total mass in the system then the position vector $\mathbf{R}$ of the centre of mass is defined by the equation

$$M\mathbf{R} = \sum_i m_i \mathbf{r}_i. \tag{5.3}$$

Proposition 5.1   *The centre of mass of the system moves in such a way that it is influenced by the external forces only and as though the total mass in the system were concentrated at it*

$$M\ddot{\mathbf{R}} = \sum_i \mathbf{F}_i.$$

Proof   On differentiating and using eqns (5.2)

$$M\ddot{\mathbf{R}} = \sum_i m_i \ddot{\mathbf{r}}_i = \sum_i \sum_{j \neq i} \mathbf{F}_{ij} + \sum_i \mathbf{F}_i$$

and (5.1) ensures that

$$M\ddot{\mathbf{R}} = \sum_i \mathbf{F}_i. \tag{5.4}$$

Corollary 5.1   *If we are considering the motion of an isolated system of particles [no external force] then the centre of mass moves with uniform velocity and can be used as the origin of a new inertial frame.*

Example

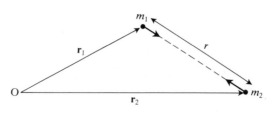

Fig. 5.2

Consider an isolated binary star system where the only force acting is the gravitational force between the two stars. Newton's second law for each particle gives

$$m_1\ddot{\mathbf{r}}_1 = \frac{Gm_1m_2}{r^2}\hat{\mathbf{r}}$$

$$m_2\ddot{\mathbf{r}}_2 = -\frac{Gm_2m_2}{r^2}\hat{\mathbf{r}}$$

where $\mathbf{r} = \mathbf{r}_2 - \mathbf{r}_1$ and $\hat{\mathbf{r}}$ is a unit vector in the same direction. On adding we see that

$$(m_1 + m_2)\ddot{\mathbf{R}} = \mathbf{0}. \tag{5.5}$$

The centre of mass must move with uniform velocity in a straight line. This means that we can use the centre of mass as the origin of a new inertial frame. Also, dividing each equation by the appropriate mass and subtracting gives

$$\ddot{\mathbf{r}} = \ddot{\mathbf{r}}_2 - \ddot{\mathbf{r}}_1 = -\frac{G(m_1 + m_2)}{r^2}\hat{\mathbf{r}}. \tag{5.6}$$

Equation (5.6) gives precisely the *central force equation* that we have already solved in Chapter 2. It is usual to measure the position vector of each star in this binary system from the centre of mass so that if we define $\mathbf{r}'_i = \mathbf{r}_i - \mathbf{R}$ then

$$\mathbf{r}'_1 = -\frac{m_2}{(m_1 + m_2)}\mathbf{r} \quad \text{and} \quad \mathbf{r}'_2 = \frac{m_1}{(m_1 + m_2)}\mathbf{r}. \tag{5.7}$$

Note that $\mathbf{r}'_i$ satisfies a very similar equation to $\mathbf{r}$.

5.2   Angular momentum and moment of force

Remember that the angular momentum of a single particle about the origin O of an inertial frame is defined to be $\mathbf{r} \wedge m\dot{\mathbf{r}}$. Then for a system of particles we have that the total angular momentum about O is

$$\mathbf{L}_O = \sum_i \mathbf{r}_i \wedge m_i\dot{\mathbf{r}}_i. \tag{5.8}$$

As in Chapter 2 we hope to make some deduction from Newton's second law about the rate of change of the total angular momentum. For that we need a definition concerning the moment of a force about a point.

Definition 5.2   If a force $\mathbf{F}$ has a line of action, one point of which has position vector $\mathbf{r}$ with respect to a point O, then the *moment* of the force about O is defined to be

$$\mathbf{r} \wedge \mathbf{F} \qquad (5.9)$$

as shown in Fig. 5.3 below.

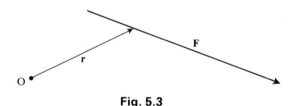

Fig. 5.3

[Note that the definition is independent of the choice of point on the line of action. To see this just consider two points on the line and note that the vector between them is parallel to the force.]

[Those readers who have used the idea of moment in a two-dimensional context will realize that this definition is a generalization. The magnitude of the moment is the magnitude of the force multiplied by the perpendicular distance to the line of action. The direction is perpendicular to the plane containing the line of action of the force and the point about which the moment is taken.]

Proposition 5.2   *For a system of particles moving in an inertial frame origin O the rate of change of the total angular momentum about O is equal to the moment of the external forces about O*

$$\frac{d\mathbf{L}_\mathbf{O}}{dt} = \sum_i \mathbf{r}_i \wedge \mathbf{F}_i \qquad (5.10)$$

provided the internal force between any two particles acts along the line joining them.

Proof   For each particle we have

$$m_i \ddot{\mathbf{r}}_i = \sum_{j \neq i} \mathbf{F}_{ij} + \mathbf{F}_i$$

$$\Rightarrow \quad \mathbf{r}_i \wedge m_i \ddot{\mathbf{r}}_i = \mathbf{r}_i \wedge \sum_{j \neq i} \mathbf{F}_{ij} + \mathbf{r}_i \wedge \mathbf{F}_i. \qquad (5.11)$$

However, consider the diagram

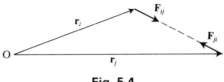

Fig. 5.4

Then

$$\mathbf{r}_i \wedge \mathbf{F}_{ij} + \mathbf{r}_j \wedge \mathbf{F}_{ji} = (\mathbf{r}_i - \mathbf{r}_j) \wedge \mathbf{F}_{ij} = 0$$

and so summing (5.11) gives

$$\sum_i \mathbf{r}_i \wedge m_i\ddot{\mathbf{r}}_i = \sum_i \mathbf{r}_i \wedge \mathbf{F}_i$$

and we have

$$\dot{\mathbf{L}}_O = \sum_i \mathbf{r}_i \wedge \mathbf{F}_i$$

which establishes (5.10).

These two results, (5.4) concerning the motion of the centre of mass and (5.10) concerning the total angular momentum, are very important and have an immediate application to the motion of a rigid body as we shall see.

Exercises:
Chapter 5

1. The only force acting upon two particles P_1 and P_2, each of mass m, is their mutual attraction which is of magnitude $\mu m r_{12}^{-2}$, where r_{12} is the distance between the particles. Initially, the particles are at a distance $2a$ apart and are moving with velocities

$$\frac{1}{2}\sqrt{\left(\frac{\mu}{a}\right)}(\sqrt{(3)}\mathbf{i} + \mathbf{j}) \quad \text{and} \quad \frac{1}{2}\sqrt{\left(\frac{3\mu}{a}\right)}(-\mathbf{i} + \sqrt{(3)}\mathbf{j})$$

respectively, where $\mathbf{i}$ is a unit vector in the direction P_1P_2 and $\mathbf{j}$ is a unit vector orthogonal to $\mathbf{i}$. Show that their centre of mass O' moves with velocity $\sqrt{(\mu/a)}\mathbf{j}$ throughout the motion and find the polar equation of the path of one of the particles relative to O'.
[Hint: use the equation $d^2u/d\theta^2 + u = -F(u^{-1})/mh^2u^2$ established in Chapter 2.]

(Oxford)

2. Consider three stars of equal mass M which move as an isolated system. Find the equation of motion for each star given that r_i is the position vector of the ith star relative to an inertial frame with origin at the centre of mass of the tertiary system. Show that the stars can move with fixed angular veolcity ω equally spaced on a circle of radius a provided that

$$\omega^2 = \frac{GM}{\sqrt{(3)a^3}}.$$

6 Rigid bodies: equations

6.1 Rigid bodies: degrees of freedom

In Newtonian mechanics a rigid body (or solid object) has a particular significance. The equations are developed from an assumption which defines the relationship between the particles which collectively form the body.

Definition 6.1 A rigid body is defined to be a collection of particles such that each particle remains in the same position relative to all other particles of the body. This ensures that the body retains its size, shape, and distribution of mass.

ASSUMPTION

The forces between any two particles obey Newton's third law (they are equal and opposite) and these internal forces act along the line joining the particles. In the notation introduced in the last chapter, $\mathbf{F}_{ij}$ is equal to $-\mathbf{F}_{ji}$ and is parallel to $\mathbf{r}_j - \mathbf{r}_i$.

A stone, a pencil, the Moon are all examples of rigid bodies which we see in everyday life.

One important feature of rigid bodies lies in the fact that we do not need to specify the position of each particle in the body since the particles are in fixed position relative to each other. It is important to know how many scalar parameters we need in order to describe the exact position of a particular body or indeed a system of particles.

Definition 6.2 The number of degrees of freedom of a mechanical system is the number of independent scalar coordinates required to fix its position precisely at a given instant in an inertial frame.

For example, the number of degrees of freedom of a particle is in general three unless some constraint is placed upon it, e.g. the bob of a spherical pendulum has only two degrees of freedom provided the string attaching it to its point of suspenison remains taut.

Proposition 6.1 *In general unconstrained motion a rigid body has six degrees of freedom.*

Proof   Suppose that G is the centre of mass of the body, as defined in Chapter 5 for a system of particles. Three Cartesian scalar variables with respect to an inertial frame will fix G. We then need to consider the orientation of the body around the centre of mass.

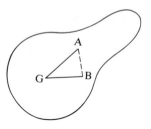

Fig. 6.1

If we locate any other point A fixed in the body we require just two further scalar variables to do so, since AG is a fixed length. Taking any third point B of the body not on the line AG we only need one further scalar variable to determine its position as B is fixed distance from both G and A. Once these three points G, A, B are located then the position of any other point of the body is known. Counting up we require six scalar variables to locate the body precisely.

From this description we can deduce that in motion the velocity of any point of the body is known if we have the velocity of the centre of mass and the angular velocity describing the rotation of the body about its centre of mass.

6.2   The angular velocity

Proposition   *In any motion of a rigid body in an inertial frame there exists an angular*
6.2   *velocity vector $\boldsymbol{\omega}$ such that if A and B are any two points fixed in the body*
                 with velocities $\mathbf{v}_A$ and $\mathbf{v}_B$ then

$$\mathbf{v}_B = \mathbf{v}_A + \boldsymbol{\omega} \wedge \mathbf{r}_{AB} \tag{6.1}$$

where $\mathbf{r}_{AB}$ is the position vector of B relative to A. The angular velocity $\boldsymbol{\omega}$ is unique and independent of A and B.

Proof   As shown in Fig. 6.2 we choose a set of orthogonal unit vectors $\mathbf{e}_i'$ fixed in direction in the body. Then we can write

$$\mathbf{r}_{AB} = \sum_i x_i \mathbf{e}_i'$$

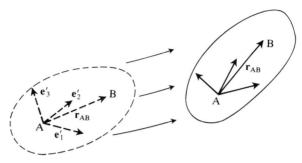

Fig. 6.2

where $x_i = \mathbf{r}_{AB} \cdot \mathbf{e}'_i$, $i = 1, 2, 3$, are constant. Differentiating in the inertial frame gives

$$\frac{d\mathbf{r}_{AB}}{dt} = \sum_i x_i \dot{\mathbf{e}}'_i.$$

The proof now follows proof 1 of the existence of the angular velocity, (p. 61), as we can establish by the identical method that

$$\boldsymbol{\omega} = \tfrac{1}{2} \sum_i \mathbf{e}'_i \wedge \dot{\mathbf{e}}'_i$$

and hence $\dot{\mathbf{e}}'_i = \boldsymbol{\omega} \wedge \mathbf{e}'_i$. Since the choice of unit vectors $\mathbf{e}'_i$ is independent of the points A and B we see that $\boldsymbol{\omega}$ is independent of the point of the body chosen. Furthermore the same argument as in Chapter 4 ensures that $\boldsymbol{\omega}$ is unique. Then using $\mathbf{r}_B - \mathbf{r}_A = \mathbf{r}_{AB}$ so that $d\mathbf{r}_{AB}/dt = \mathbf{v}_B - \mathbf{v}_A$, the result follows.

The only case where we have an exception to the proposition above is in the rather unphysical case of the infinitely thin rod. In this instance the angular velocity is only determined uniquely perpendicular to the rod and has an arbitrary component along the rod.

Definition 6.3   If a body is moving in an inertial frame in such a way that there is a line of points in the body which is instantaneously at rest then that line is called the instantaneous axis of rotation.

By using eqn (6.1) the velocity of any point P of the body can be determined if $\mathbf{v}_G$, the velocity of the centre of mass, and the angluar velocity are known since

$$\mathbf{v}_P = \mathbf{v}_G + \boldsymbol{\omega} \wedge \mathbf{r} \tag{6.2}$$

where $\mathbf{r}$ is the vector from G to P.

Proposition 6.3   *An instantaneous axis of rotation exists if and only if $\mathbf{v}_G \cdot \boldsymbol{\omega} = 0$.*

Proof   We are looking for points which instantaneously have zero velocity. Using (6.2)

$$0 = \mathbf{v}_G + \boldsymbol{\omega} \wedge \mathbf{r}.$$

This vector equation has solution for $\mathbf{r}$ if and only if $\mathbf{v}_G.\boldsymbol{\omega} = 0$. If this condition is satisfied then

$$\mathbf{r} = -\frac{\mathbf{v}_G \wedge \boldsymbol{\omega}}{\omega^2} + \lambda\boldsymbol{\omega} \qquad (6.3)$$

giving a line of points parallel to $\boldsymbol{\omega}$.

NB It follows that if one point of the body is anchored in the inertial frame then the instantaneous axis must exist as a line passing through that point parallel to $\boldsymbol{\omega}$.

6.3   The equations of motion

Given that in general a rigid body has six degrees of freedom the most important equation developed so far is eqn (6.2) which we now label

(RB1) $\mathbf{v}_P = \mathbf{v}_G + \boldsymbol{\omega} \wedge \mathbf{r}$

where $\mathbf{r} = \underline{GP}$. In the rest of this chapter $\mathbf{v}$ will refer to the velocity of the centre of mass, dropping the subscript G. If we can find equations from Newton's second law which develop first-order differential equations for the linear and angular velocities $\mathbf{v}$ and $\boldsymbol{\omega}$ then in theory we can solve any problem involving motion of a rigid body. We will have produced a comprehensive model for any such motion under mechanical forces. The systems of many particles which we dealt with in Chapter 5 give us a clue as to how to set about finding such equations. All the following propositions hold in any inertial frame.

LINEAR MOMENTUM

Proposition 6.4   *If M is the total mass of the rigid body and $\mathbf{F}_i$ represent the external forces acting on it then the rate of change of the linear momentum is the vector sum of the external forces*

(RB2) $$M\frac{d\mathbf{v}}{dt} = \sum \mathbf{F}_i \qquad (6.4)$$

where $\mathbf{v}$ is the velocity of the centre of mass.

Proof

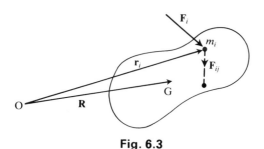

Fig. 6.3

Considering the body as a collection of particles subject to both internal and external forces, then using the assumption that internal forces are equal and opposite between pairs of particles we have

$$m_i\ddot{\mathbf{r}}_i = \sum_{j \neq i} \mathbf{F}_{ij} + \mathbf{F}_i \qquad \mathbf{F}_{ij} = -\mathbf{F}_{ji}$$

with respect to some chosen inertial frame. On adding and using the definition of **R**, the position vector of the centre of mass,

$$M\mathbf{R} = \sum m_i \mathbf{r}_i$$

$$\sum m_i \ddot{\mathbf{r}}_i = \sum \mathbf{F}_i \Rightarrow M\ddot{\mathbf{R}} = \sum \mathbf{F}_i.$$

Since $\mathbf{v} = \dot{\mathbf{R}}$ this proves the result.

ANGULAR MOMENTUM

Looking at the previous result for systems of particles we need to use moments of forces. So for completeness we repeat the definition required.

Definition 6.4   The moment of a force **F** about a point O is defined to be

$$\mathbf{r} \wedge \mathbf{F}$$

where **r** is any position vector from O to the line of action of the force.

Proposition 6.5   *The rate of change of angular momentum about the centre of mass G of a rigid body is the moment of the external forces about G*

(RB3)

$$\frac{d\mathbf{L}_G}{dt} = \sum \mathbf{r}_i' \wedge \mathbf{F}_i \qquad (6.5)$$

where $\mathbf{r}_i'$ is a position vector from G to the line of action of the force $\mathbf{F}_i$.

Proof

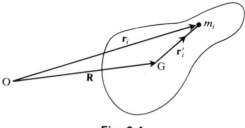

Fig. 6.4

Taking a vector product of Newton's second law for an individual particle with the vector from G to the particle we have

$$m_i(\mathbf{r}_i - \mathbf{R}) \wedge \ddot{\mathbf{r}}_i = \sum_j (\mathbf{r}_i - \mathbf{R}) \wedge \mathbf{F}_{ij} + (\mathbf{r}_i - \mathbf{R}) \wedge \mathbf{F}_i.$$

From the assumption that $\mathbf{F}_{ij}$ is equal to $-\mathbf{F}_{ji}$ and is parallel to the vector $\mathbf{r}_j - \mathbf{r}_i$, we can see that summing over all particles the moment of the internal forces disappears.

Hence

$$\sum m_i(\mathbf{r}_i - \mathbf{R}) \wedge \ddot{\mathbf{r}}_i = \sum (\mathbf{r}_i - \mathbf{R}) \wedge \mathbf{F}_i. \tag{6.6}$$

It remains to link the left-hand side of this equation to the angular momentum:

$$\mathbf{L}_G = \sum m_i(\mathbf{r}_i - \mathbf{R}) \wedge \dot{\mathbf{r}}_i.$$

Differentiating gives

$$\dot{\mathbf{L}}_G = \sum m_i(\dot{\mathbf{r}}_i - \dot{\mathbf{R}}) \wedge \dot{\mathbf{r}}_i + \sum m_i(\mathbf{r}_i - \mathbf{R}) \wedge \ddot{\mathbf{r}}_i$$
$$= -\dot{\mathbf{R}} \wedge \sum m_i \dot{\mathbf{r}}_i + \sum m_i(\mathbf{r}_i - \mathbf{R}) \wedge \ddot{\mathbf{r}}_i.$$

However, since $M\dot{\mathbf{R}} = \sum m_i \dot{\mathbf{r}}_i$ the first expression on the right is $\mathbf{0}$.

Hence from (6.6) we have

$$\dot{\mathbf{L}}_G = \sum (\mathbf{r}_i - \mathbf{R}) \wedge \mathbf{F}_i$$

giving the required result.

These two equations, (RB2) and (RB3), are precisely what is needed to provide a model for rigid body motion if we can make the connection between the angular momentum and the angular velocity of the body. That there is a connection is plausible if we consider the angular momentum of a particle moving in a plane. Then $\mathbf{L}_O = mr^2\dot{\theta}\mathbf{n}$ where $\mathbf{n}$ is the normal to the plane of motion and r, θ are polar coordinates with origin O. Note that $\dot{\theta}\mathbf{n}$ is the angular velocity of the particle about O.

Considering the situation for a rigid body which is in equilibrium, then the equations involving linear and angular momentum produce the following corollary.

Corollary   *If a rigid body is in equilibrium then the vector sum of the external forces is zero and the moment of the forces about each point P is also zero, at all times t.*

Proof

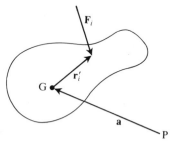

Fig. 6.5

Since the velocity of the centre of mass is zero then $\dot{\mathbf{v}} = \mathbf{0}$

$$\Rightarrow \quad \sum \mathbf{F}_i = \mathbf{0}.$$

Since $\ddot{\mathbf{r}}_i = \mathbf{0}$ for every particle then $\dot{\mathbf{L}}_G = \mathbf{0}$ also,

$$\Rightarrow \quad \sum \mathbf{r}'_i \wedge \mathbf{F}_i = \mathbf{0}.$$

Hence the moment of the forces about the centre of mass is zero.

About any other point P, if $\Gamma(\mathbf{P})$ denotes the total moment about P and $\mathbf{a} = \underline{\mathrm{PG}}$,

$$\Gamma(\mathbf{P}) = \sum (\mathbf{a} + \mathbf{r}'_i) \wedge \mathbf{F}_i$$
$$= \mathbf{a} \wedge \sum \mathbf{F}_i + \sum \mathbf{r}'_i \wedge \mathbf{F}_i$$
$$= \mathbf{0}.$$

It is clear, on reversing the above argument, that if the moment about P is zero then the moment about G is also zero, given that the vector sum of the forces is zero.

There is one case where the equations (RB2) and (RB3) can be replaced by a single equation, giving a considerable simplification. That occurs when the body rotates about one point fixed both in the body and in the inertial frame, rotation about an anchored point.

Proposition 6.6   *If the body is rotating about one point Q of itself which is anchored in an inertial frame, then the rate of change of the anglar momentum about Q is equal to the moment of the external forces about Q*

(RB3′)
$$\dot{\mathbf{L}}_Q = \sum \mathbf{s}_i \wedge \mathbf{F}_i \tag{6.7}$$

where $\mathbf{s}_i$ is a position vector from Q to the line of action of the force $\mathbf{F}_i$.

The proof is exactly similar and rather easier than the proof for the general case of rate of change of angular momentum about the centre of mass and is left to the reader.

It now remains to establish the connection between the angular velocity of the body and the angular momentum about the centre of mass.

6.4   Angular momemtum and angular velocity

A simple example gives us a clue as to how important this connection turns out to be. Suppose that you incline a table so that its surface is at a small angle to the horizontal. Take two identical beer cans. Empty one in whatever way seems most appropriate. Then roll them side by side down lines of greatest slope releasing the two cans from rest at the same moment. You need to make sure that the cans roll and do not slip across the surface. Mathematically if we attempted to treat the two cans as particles sliding down a smooth surface the mass of the cans would make no difference to the time it takes to roll to the bottom of the table. You can test this statement by taking two full cans of different mass. To within experimental error they will take the same time to roll down the table. However in the case of the empty can and the full can one of them takes a substantially longer time to get to the bottom. What is important here is the distribution of the mass. In some way there is a connection between the distribution of mass, the angular velocity, and the angular momentum of the can.

We start from the definition of angular momentum for a particle

$$\mathbf{L}_O = \mathbf{r} \wedge m\mathbf{v}$$

where $\mathbf{r}$ is the position vector of the particle relative to O and $\mathbf{v}$ is the true velocity of the particle in an inertial frame. The most important point in a rigid body and the point about which we can think of the body rotating is the centre of mass.

Definition 6.5   The angular momentum of a rigid body about the centre of mass is

$$\mathbf{L}_G = \sum \mathbf{r} \wedge m\mathbf{v}'$$

where the sum is taken over all the particles which constitute the body, $\mathbf{r}$ is the position vector of the particle from the centre of mass, and $\mathbf{v}'$ is the velocity of the particle in the inertial frame in which the body is moving.

Proposition 6.7   *If axes with unit vectors $\{\mathbf{e}_i\}$ are chosen fixed in the body and $\mathbf{L}_G$, $\boldsymbol{\omega}$ are written as column vectors with respect to these axes then there exists a symmetric matrix $\mathfrak{I}_G$ such that*

(RB4)
$$\mathbf{L}_G = \mathfrak{I}_G \boldsymbol{\omega} \tag{6.8}$$

where $\boldsymbol{\omega}$ is the angular velocity of the body. $\mathfrak{I}_G$ depends only on the geometry of the body and its distribution of mass.

Proof

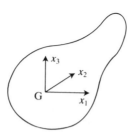

Fig. 6.6

Select a set of axes at G, say $Gx_1x_2x_3$ with unit vectors $\{\mathbf{e}_i\}$ which are fixed in the body and therefore rotate with it with the same angular velocity $\boldsymbol{\omega}$, with respect to the inertial frame in which the body moves.

Firstly in vectors we have

$$\mathbf{L}_G = \sum \mathbf{r} \wedge m\mathbf{v}'$$
$$= \sum \mathbf{r} \wedge m(\mathbf{v} + \boldsymbol{\omega} \wedge \mathbf{r}) \quad \text{using (RB1) (6.2)}$$
$$= \sum \mathbf{r} \wedge m\mathbf{v} + \sum m\mathbf{r} \wedge (\boldsymbol{\omega} \wedge \mathbf{r})$$

where $\mathbf{v}$ is the velocity of the centre of mass. Since $\mathbf{r}$ is measured from the centre of mass

$$\sum m\mathbf{r} = 0 \Rightarrow \sum m\mathbf{r} \wedge \mathbf{v} = 0$$

giving

$$\mathbf{L}_G = -\sum m\mathbf{r} \wedge (\mathbf{r} \wedge \boldsymbol{\omega}).$$

In order to sort out the remaining term notice that we can represent the vector product of two vectors in the following way using components:

$$\mathbf{a} \wedge \mathbf{b} = \begin{bmatrix} 0 & -a_3 & a_2 \\ a_3 & 0 & -a_1 \\ -a_2 & a_1 & 0 \end{bmatrix} \begin{bmatrix} b_1 \\ b_2 \\ b_3 \end{bmatrix}.$$

If $\mathbf{r} = \Sigma x_i e_i$ then

$$
\mathbf{L}_G = -\Sigma m \begin{bmatrix} 0 & -x_3 & x_2 \\ x_3 & 0 & -x_1 \\ -x_2 & x_1 & 0 \end{bmatrix}^2 \begin{bmatrix} \omega_1 \\ \omega_2 \\ \omega_3 \end{bmatrix}
$$

$$
= \begin{bmatrix} A & -H & -G \\ -H & B & -F \\ -G & -F & C \end{bmatrix} \begin{bmatrix} \omega_1 \\ \omega_2 \\ \omega_3 \end{bmatrix}
$$

$$
= \mathfrak{I}_G \omega
$$

where $A = \Sigma m(x_2^2 + x_3^2)$, $F = \Sigma mx_2 x_3$, etc. The proposition is proved.

The proof calls for a further definition and an explanation of what precisely the matrix $\mathfrak{I}_G$ entails.

Definition 6.6
The inertia tensor at G with respect to the axes $Gx_1 x_2 x_3$ is defined to be

$$
\mathfrak{I}_G = \begin{bmatrix} A & -H & -G \\ -H & B & -F \\ -G & -F & C \end{bmatrix} \tag{6.9}
$$

where the moments of inertia are $A = \Sigma m(x_2^2 + x_3^2)$, $B = \Sigma m(x_3^2 + x_1^2)$, $C = \Sigma m(x_1^2 + x_2^2)$, and the products of inertia are F, G, H with $F = \Sigma mx_2 x_3$, $G = \Sigma mx_3 x_1$, and $H = \Sigma mx_1 x_2$.

In practice of course the moments and products of inertia are found using multiple integrals. We can also find the inertia tensor at any point Q in the body although a general point Q will not normally give us such a convenient relationship between the angular velocity and the angular momentum. There is one case where it is known that there is a similar relationship, however.

Proposition 6.8
(RB4′)
If the body is rotating about one point Q which is fixed ($\mathbf{v}_Q = \mathbf{0}$) then

$$
\mathbf{L}_Q = \mathfrak{I}_Q \omega \tag{6.10}
$$

where the inertia tensor is evaluated at a set of axes fixed in the body at Q.

The proof is left to the reader since it follows the same lines as the proof for $\mathbf{L}_G$. Note that since $\mathbf{v}_Q = \mathbf{0}$ then the velocity of a particle of the body $\mathbf{v}_p$ is given by

$$
\mathbf{v}_p = \omega \wedge \mathbf{s}_p
$$

where $\mathbf{s}_p$ is the position vector of the particle from Q.

To complete the picture with regard to angular momentum it may be that we will need to calculate the angular momentum about some point P other than the centre of mass for which $\mathbf{v_P} \neq \mathbf{0}$. The following proposition fills this last gap.

Proposition 6.9

If the centre of mass G has position vector $\mathbf{s}$ *measured from origin* P *which is a general point fixed in the body (and not necessarily fixed in the inertial frame), then the angular momentum of the body about* P *is related to that about* G *by the equation*

(RB5)

$$\mathbf{L_P} = \mathbf{s} \wedge M\mathbf{v} + \mathbf{L_G} \qquad (6.11)$$

where M is the mass of the body and $\mathbf{v}$ *is the velocity of the centre of mass.*

In words: the angular momentum about P *is the angular momentum about* G *together with the angular moment about* P *of a particle of equivalent mass situated at* G.

Proof

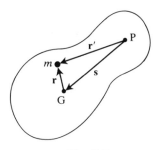

Fig. 6.7

$$\mathbf{L_P} = \sum \mathbf{r'} \wedge m\mathbf{v'}$$

where $\mathbf{r'}$ is measured from P and $\mathbf{v'}$ is the true velocity of the individual particle

$$\mathbf{L_P} = \sum (\mathbf{s} + \mathbf{r}) \wedge m\mathbf{v'}$$
$$= \mathbf{s} \wedge \sum m\mathbf{v'} + \mathbf{L_G}$$
$$= \mathbf{s} \wedge M\mathbf{v} + \mathbf{L_G}$$

since $\Sigma\, m\mathbf{v'} = M\mathbf{v}$ from the definition of the centre of mass.

Example

Treat the rolling can as a cylinder uniform along its length. The only other assumption about the mass must be that it is distributed symmetrically about the axis of symmetry of the can. Essentially the motion is two-dimensional so that we need only consider the vertical plane through a line of greatest slope of the inclined table and the centre of mass of the cylinder.

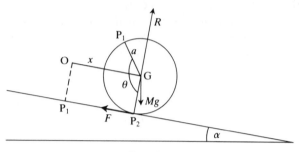

Fig. 6.8

The cylinder is rolling without slipping. If x is the distance moved down the slope and θ is the angle turned through by the cylinder in the same time then

$$x = a\theta.$$

Using (RB2), the equation for linear momentum down the slope, we have

$$M\ddot{x} = -F + Mg \sin \alpha$$

where F is the frictional force. Using (RB3), the equation for the angular momentum about G, in the direction of the axis of symmetry of the cylinder, we have

$$I\ddot{\theta} = aF$$

where I is the moment of inertia about this axis

$$\Rightarrow \quad M\ddot{x} = -I\frac{\ddot{x}}{a^2} + Mg \sin \alpha$$

$$\Rightarrow \quad \ddot{x} = \frac{Ma^2}{(I + Ma^2)} g \sin \alpha.$$

Now we can see that the larger the ratio I/Ma^2, the smaller the acceleration down the plane and the longer the can takes to roll to the bottom. The hollow cylinder must have a larger ratio than the solid one.

Important
Those students who have studied neither multiple integrals nor symmetric matrices can ignore most of the following two sections with the exception of

a practial summary at the end of Section 6.6. This covers special cases of bodies with some rotational symmetry and gives principal moments of inertia for well-known shapes.

6.5   Calculation of the inertia tensor

Calculating the moments and products of inertia for a specific set of axes involves the geometry of the body, knowing how the mass is distributed, the density, and finally multiple intergrals.

We start by considering some easy examples.

Example 1   A lamina is in the shape of a thin disc with constant surface density σ per unit area and radius a. Choose axes Gz perpendicular to the disc through its centre of mass and Gx, Gy any two perpendicular axes in the plane of the disc.

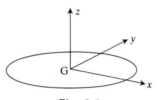

Fig. 6.9

With respect to the Cartesian axes any point on the disc has coordinates of the form $(x, y, 0)$.

Treating the moments of inertia first

$$C = \sum m(x^2 + y^2) = \int \int \sigma(x^2 + y^2)dx\,dy$$

on replacing $m \approx \sigma\,dx\,dy$. Changing to polar coordinates

$$C = \int_{r=0}^{a} \int_{\theta=0}^{2\pi} \sigma r^3\,dr\,d\theta$$

$$= \tfrac{1}{2}Ma^2$$

where $M = \pi a^2 \sigma$. By a symmetry arugment $A = B$ and

$$A = \sum m(y^2 + 0) = \int \int \sigma y^2\,dx\,dy = \tfrac{1}{2}C = \tfrac{1}{4}Ma^2.$$

To find the products of inertia note that since $z = 0$ for every point of the disc then $F = G = 0$, and

$$H = \sum mxy = \iint \sigma xy \, dx \, dy = \int_{r=0}^{a} \int_{\theta=0}^{2\pi} \sigma r^3 \sin \theta \cos \theta \, dr \, d\theta = 0.$$

We could deduce this result from a symmetry argument since if the point $(x, y, 0)$ lies on the disc all the points $(\pm x, \pm y, 0)$ also lie on the disc. Hence

$$\mathfrak{I}_G = \begin{bmatrix} \frac{1}{4}Ma & 0 & 0 \\ 0 & \frac{1}{4}Ma^2 & 0 \\ 0 & 0 & \frac{1}{2}Ma^2 \end{bmatrix}.$$

Example 2   Now consider a solid uniform hemisphere with radius a and mass $M = \frac{2}{3}\pi\rho a^3$. First choose axes at P the midpoint of the base, as shown.

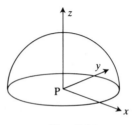

Fig. 6.10

Again $C = \sum m(x^2 + y^2)$ and we replace the sum by integration using $m \approx \rho \, dx \, dy \, dz$:

$$C = \iiint \rho(x^2 + y^2)dx \, dy \, dz.$$

Using spherical polars

$$C = \int_{r=0}^{a} \int_{\theta=0}^{\pi/2} \int_{\phi=0}^{2\pi} \rho(r^2 \sin^2 \theta)r^2 \sin \theta \, dr \, d\theta \, d\phi$$

$$= \frac{2}{5}Ma^2.$$

By symmetry $A = B$ and

$$B = \iiint \rho(r^2 \sin^2 \theta \cos^2 \phi + r^2 \cos^2 \theta)r^2 \sin \theta \, dr \, d\theta \, d\phi$$

with the same limits $\Rightarrow B = \frac{2}{5}Ma^2$.

It is easy to see by a symmetry argument that $F = G = H = 0$ giving

$$\mathfrak{I}_P = \tfrac{2}{5}Ma^2 I$$

where I is the identity matrix. We have found the inertia tensor at the midpoint of the base not at the centre of mass.

From the definition of the centre of mass we can deduce that it must lie on the axis of symmetry of the hemisphere. If $\bar{z}$ is the distance above P

$$M\bar{z} = \iiint \rho z \, dx \, dy \, dz$$

$$= \int_{r=0}^{a} \int_{\theta=0}^{\pi/2} \int_{\phi=0}^{2\pi} \rho r \cos \theta r^2 \sin \theta \, dr \, d\theta \, d\phi$$

$$\bar{z} = \tfrac{3}{8}a.$$

How do we find $\mathfrak{I}_G$ knowing the position of the centre of mass and $\mathfrak{I}_P$?

Proposition 6.10 (*The parallel axes theorem*) *If* G *is the centre of mass and* P *any other point of the body then*

$$\mathfrak{I}_P = \mathfrak{I}_G + M(\mathbf{p}^{\mathsf{T}}\mathbf{p}I - \mathbf{p}\mathbf{p}^{\mathsf{T}}) \tag{6.12}$$

where **p** *is the column vector representing the vector from* G *to* P *with respect to the chosen axes at* G *and* $\mathfrak{I}_P$ *is determined with respect to a set of parallel axes at* P. *The mass of the body is* M.

Proof

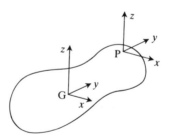

Fig. 6.11

The proof is probably easiest to understand using a simple approach, although you can derive a neat version of it using the matrix form of the vector product $(\mathbf{r} \wedge \quad)$ as in the derivation of the expression for the angular momentum.

Suppose that $\underline{GP} = a\mathbf{i} + b\mathbf{j} + c\mathbf{k}.$

Looking first at moments of inertia

$$A_P = \sum m[(y - b)^2 + (z - c)^2]$$
$$= A_G - 2b \sum my - 2c \sum mz + M(b^2 + c^2).$$

By definition of the centre of mass the two middle terms are 0. Hence

$$A_P = A_G + M(b^2 + c^2)$$

and similar results follow for B, C.

For the products of inertia

$$F_P = \sum m(y - b)(z - c)$$
$$= F_G + Mbc$$

and similar results follow for G, H.

Hence

$$\Im_P = \Im_G + \begin{bmatrix} M(b^2 + c^2) & -Mab & -Mca \\ -Mab & M(c^2 + a^2) & -Mbc \\ -Mca & -Mbc & M(a^2 + b^2) \end{bmatrix}$$

and it is easy to check that if $\mathbf{p}^T = (a, b, c)$ then the last matrix is neatly written as

$$M(\mathbf{p}^T\mathbf{p}I - \mathbf{p}\mathbf{p}^T)$$

proving the proposition.

A word of caution: *the theorem only works between the centre of mass and another point.*

Example 2 (cont.)   For the hemisphere $\mathbf{p}^T = (0, 0, \frac{3}{8}a)$ and

$$\Im_P = \tfrac{2}{5}Ma^2 I = \Im_G + M \left(\tfrac{9}{64}a^2 I - \begin{bmatrix} 0 & 0 & 0 \\ 0 & 0 & 0 \\ 0 & 0 & \frac{9}{64}a^2 \end{bmatrix} \right).$$

This gives $\Im_G$ as a diagonal matrix whose first two diagonal entries are $(\frac{2}{5} - \frac{9}{64})Ma^2$ and the moment of inertia along the axis of symmtery is unchanged at $\frac{2}{5}Ma^2$.

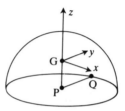

Fig. 6.12

If we wish to do so we can now find the inertia tensor at a point on the rim, say

$$\underline{GQ} = \mathbf{q}^{\mathrm{T}} = (0, a, -\tfrac{3}{8}a).$$

Then

$$\mathfrak{I}_Q = \mathfrak{I}_G + M\left(\tfrac{73}{64}a^2 I - \begin{bmatrix} 0 & 0 & 0 \\ 0 & a^2 & -\tfrac{3}{8}a^2 \\ 0 & -\tfrac{3}{8}a^2 & \tfrac{9}{64}a^2 \end{bmatrix}\right)$$

Giving

$$\mathfrak{I}_Q = \begin{bmatrix} \tfrac{7}{5}Ma^2 & 0 & 0 \\ 0 & \tfrac{2}{5}Ma^2 & \tfrac{3}{8}Ma^2 \\ 0 & \tfrac{3}{8}Ma^2 & \tfrac{7}{5}Ma^2 \end{bmatrix}.$$

You will notice that the inertia tensor is a symmetric matrix. Hence we know that a set of axes with respect to which the inertia tensor is diagonal always exists.

Proposition 6.11 *With respect to any point P of the body there exists a set of axes origin P such that the inertia tensor is diagonal.*

Proof Take an arbitrary set of axes at P and suppose that the inertia tensor is $\mathfrak{I}_P$. If we choose a new set of axes so that any column vector $\mathbf{r}$ becomes $\mathbf{r}' = W\mathbf{r}$ in the new frame, then W is a rotation matrix and must be orthogonal, $W^{\mathrm{T}}W = I$. Since $\mathfrak{I}_P$ is a matrix, with respect to the new axes we have

$$\mathfrak{I}'_P = W\mathfrak{I}_P W^{\mathrm{T}}.$$

Because $\mathfrak{I}_P$ is a symmetric matrix we know that there exists a rotation matrix U such that $U\mathfrak{I}_P U^{\mathrm{T}}$ is a diagonal matrix and $U^{\mathrm{T}}U = I$. The new axes are the mutually perpendicular eigenvectors of $\mathfrak{I}_P$ and the eigenvalues are the diagonal entries of $U\mathfrak{I}_P U^{\mathrm{T}}$.

Definition 6.7 The principal axes of inertia at P are the axes with respect to which the inertia tensor is diagonal and the principal moments of inertia are those associated with these axes.

Corllary *It follows that the eigenvalues are the principal moments of inertia and the eigenvectors are the principal axes. Furthermore these are the axes for which the corresponding products of inertia are all zero.*

Example 1 (cont.) Find the principal moments of inertia and principal axes of the inertia tensor at the point Q on the disc shown in Fig. 6.13 below.

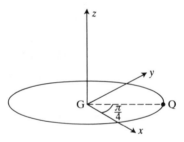

Fig. 6.13

Choosing Q to be the point on the rim of the disc $(a/\sqrt{2}, a/\sqrt{2}, 0)$ then the inertia tensor at Q becomes

$$\mathfrak{I}_Q = \begin{bmatrix} \tfrac{3}{4}Ma^2 & -\tfrac{1}{2}Ma^2 & 0 \\ -\tfrac{1}{2}Ma^2 & \tfrac{3}{4}Ma^2 & 0 \\ 0 & 0 & \tfrac{3}{2}Ma^2 \end{bmatrix}.$$

We must find the eigenvalues and eigenvectors of $\mathfrak{I}_Q$ by solving $\det(\mathfrak{I}_Q - \lambda I) = 0, (\mathfrak{I}_Q - \lambda I)\mathbf{x} = 0$:

$$\det \begin{bmatrix} \tfrac{3}{4}Ma^2 - \lambda & -\tfrac{1}{2}Ma^2 & 0 \\ -\tfrac{1}{2}Ma^2 & \tfrac{3}{4}Ma^2 - \lambda & 0 \\ 0 & 0 & \tfrac{3}{2}Ma^2 - \lambda \end{bmatrix} = 0.$$

It is easy to see that $\lambda = \tfrac{3}{2}Ma^2$ is an eigenvalue with eigenvector $(0, 0, 1)^T$. On setting $\lambda = \mu Ma^2$, this leaves us with the determinant

$$\begin{vmatrix} \tfrac{3}{4} - \mu & -\tfrac{1}{2} \\ -\tfrac{1}{2} & \tfrac{3}{4} - \mu \end{vmatrix} = 0$$

so that $\lambda = \mu Ma^2 = \tfrac{5}{4}Ma^2, \tfrac{1}{4}Ma^2$, and the corresponding eigenvectors are $(1, -1, 0)^T$ and $(1, 1, 0)^T$. These give all the principal moments and principal axes at Q. One eigenvector $(0, 0, 1)^T$ is again perpendicular to the disc; the other two are along the tangent and radius at Q respectively.

6.6   Application to calculation of angular momentum

The methods developed in the previous two sections for calculating angular momentum are best understood by means of a simple example.

Example 1 (cont.)   Suppose that a disc is rotating with angular speed ω about an axis tangential to the disc at Q as shown in the diagram. Find (i) $\mathbf{L}_Q$ and (ii) $\mathbf{L}_P$, P as below.

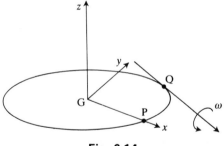

Fig. 6.14

With respect to the axes shown

$$\boldsymbol{\omega} = \left(\frac{\omega}{\sqrt{2}} \quad -\frac{\omega}{\sqrt{2}} \quad 0 \right)^{\mathrm{T}}.$$

Equation (RB4)$\Rightarrow$

$$\mathbf{L}_G = \mathfrak{I}_G \boldsymbol{\omega} = \frac{\omega}{\sqrt{2}} \begin{bmatrix} \frac{1}{4}Ma^2 & 0 & 0 \\ 0 & \frac{1}{4}Ma^2 & 0 \\ 0 & 0 & \frac{1}{2}Ma^2 \end{bmatrix} \begin{bmatrix} 1 \\ -1 \\ 0 \end{bmatrix}$$

$$\mathbf{L}_G = \frac{Ma^2\omega}{4\sqrt{2}} \begin{bmatrix} 1 \\ -1 \\ 0 \end{bmatrix}.$$

Then (RB4')$\Rightarrow$

$$\mathbf{L}_Q = \mathfrak{I}_Q \boldsymbol{\omega} = \frac{Ma^2\omega}{\sqrt{2}} \begin{bmatrix} \frac{3}{4} & -\frac{1}{2} & 0 \\ -\frac{1}{2} & \frac{3}{4} & 0 \\ 0 & 0 & \frac{1}{2} \end{bmatrix} \begin{bmatrix} 1 \\ -1 \\ 0 \end{bmatrix}$$

$$\mathbf{L}_Q = \frac{5Ma^2\omega}{4\sqrt{2}} (1 \quad -1 \quad 0)^{\mathrm{T}}.$$

This result may have been expected since we had already shown in Section 6.6 that the tangent to the disc at Q was a principal axis. Here we simply have the angular velocity multipled by the appropriate principal moment of inertia at Q.

Suppose that we now wish to calulate the angular momentum at P, $(1,0,0)$, which is neither the centre of mass nor a fixed point. Then we must use (RB5), the rule which connects the angular momentum of the centre of mass with that of any other point of the body.

Equation (RB5)⇒

$$\mathbf{L_P} = \mathbf{L_G} + \underline{PG} \wedge M\mathbf{v_G}$$

Now using (RB1)

$$\mathbf{v_G} = \mathbf{v_Q} + \boldsymbol{\omega} \wedge \underline{QG}$$

$$= \frac{\omega}{\sqrt{2}}(1 \quad -1 \quad 0) \wedge \frac{a}{\sqrt{2}}(-1 \quad -1 \quad 0)$$

$$= \frac{a\omega}{2}(0 \quad 0 \quad -2) = a\omega(0 \quad 0 \quad -1)$$

$$\mathbf{L_P} = \frac{Ma^2\omega}{4\sqrt{2}}(1 \quad -1 \quad 0) + Ma^2\omega(-1 \quad 0 \quad 0) \wedge (0 \quad 0 \quad -1)$$

$$= \frac{Ma^2\omega}{4\sqrt{2}}(1 \quad -1 \quad 0) + Ma^2\omega(0 \quad -1 \quad 0)$$

NB⇒

$$\neq \Im_P \omega.$$

This completes the demonstration of the use of the formulae for the angular momentum.

SUMMARY OF RESULTS FOR SOME WELL-KNOWN OBJECTS

1. *Regular shapes*

(a) A solid uniform sphere, radius a, mass M, has moment of inertia $\frac{2}{5}Ma^2$ about any axis through its centre G. The angular momentum about G is $\frac{2}{5}Ma^2\omega$.
(b) A hollow uniform sphere has moment of inertia $\frac{2}{3}Ma^2$ about each axis through G and $\mathbf{L_G} = \frac{2}{3}Ma^2\omega$.
(c) A uniform cube of side $2a$ has moment of inertia $\frac{2}{3}Ma^2$ about any axis through G. Again $\mathbf{L_G} = \frac{2}{3}Ma^2\omega$.

2. *Bodies with rotational symmetry about an axis*

Here we can use a general result for a body with uniform density and axial symmetry. The inertia tensor at any point P on the axis of symmetry is diagonal with respect to any two mutually perpendicular axes at P, each perpendicular to the axis of symmetry, the third axis being the axis of symmetry itself. Consequently if $\boldsymbol{\omega} = (\omega_1, \omega_2, \omega_3)$ then $\mathbf{L_P} = (A\omega_1, A\omega_2, C\omega_3)$ where A, A, C are the principal moments of inertia long these axes.

(a) A flat disc with uniform surface density and total mass M, radius a, has $A = \frac{1}{4}Ma^2$ and $C = \frac{1}{2}Ma^2$ at its centre. The disc is symmetrical about the axis through its centre normal to the surface.
(b) A circular ring of uniform line density, mass M, and radius a has moments $A = \frac{1}{2}Ma^2$ and $C = Ma^2$ about its centre.
(c) A uniform solid cylinder of radius a and height $2h$ has moments $A = M(\frac{1}{4}a^2 + \frac{1}{3}h^2)$ and $C = \frac{1}{2}Ma^2$ about its centre of mass.
(d) A thin rod of uniform line density and length $2a$ has moment $A = \frac{1}{3}Ma^2$ about any axis perpendicular to itself through its centre of mass and zero moment about the axis along its length.

6.7   Energy

As was the case of our study of particle mechanics the one remaining concept is that of energy and in particular kinetic energy, since the treatment of potential energy depends only on the external forces and follows exactly the same argument as for a particle. Potential energy exists and energy is conserved if the external foces involved are conservative and/or are reactions which do no work in the motion of the body.

Proposition 6.12

(RB6)

If a rigid body is in general motion its kinetic energy can be expressed as

$$T = \tfrac{1}{2}Mv^2 + \tfrac{1}{2}\omega.L_G \qquad (6.13)$$

where v is the velocity of the centre of mass, ω is the angular velocity, and L_G is the angular momentum about the centre of mass.

[This expression indicates that, as we might expect, the kinetic energy is made up from two parts: one the linear momentum it would have if it were modelled as a single particle, the second the rotational energy about the centre of mass.]

Proof

$$T = \tfrac{1}{2}\sum mv'.v'$$

where v' is the velocity of the particle and the sum is taken over the constituent particles.
Thus using (RB1) to replace the second v' but not the first,

$$T = \tfrac{1}{2}\sum mv'.(v + \omega \wedge r')$$
$$= \tfrac{1}{2}\sum mv'.v + \tfrac{1}{2}\sum mv'.\omega \wedge r'$$

where $\mathbf{r}'$ is the position vector of a particle from G. Using the definition of the centre of mass for the first term, $M\mathbf{v} = \Sigma \, m\mathbf{v}'$, and the properties of the triple scalar product for the second term we have

$$T = \tfrac{1}{2}M\mathbf{v}.\mathbf{v} + \tfrac{1}{2}\boldsymbol{\omega}.\textstyle\sum mr' \wedge v'$$
$$= \tfrac{1}{2}M\mathbf{v}.\mathbf{v} + \tfrac{1}{2}\boldsymbol{\omega}.\mathbf{L}_G$$

completing the proof.

As usual there is one special case in which it is easy to caluclate the kinetic energy, that is when the body is rotating about a point Q which is anchored in an inertial frame.

Proposition 6.13

If a body is rotating about a point Q of itself which is anchored in an inertial frame then the kinetic energy is given by

(RB6′)

$$T = \tfrac{1}{2}\boldsymbol{\omega}.\mathbf{L}_Q. \tag{6.14}$$

Proof

Since $\mathbf{v}_Q = \mathbf{0}$ we have

$$T = \tfrac{1}{2}\sum m\mathbf{v}'.\mathbf{v}'$$
$$= \tfrac{1}{2}\sum m\mathbf{v}'.(\boldsymbol{\omega} \wedge \mathbf{r}'')$$

where $\mathbf{r}''$ is the position vector of a particle from Q.

On rearranging the triple scalar product as before we have the required result.

Example 1 (cont.)

Consider the disc rotating as before about a point Q on the rim with angular velocity directed along the tangent at Q.

Using (RB6) we have

$$T = \tfrac{1}{2}M\mathbf{v}.\mathbf{v} + \tfrac{1}{2}\boldsymbol{\omega}.\mathbf{L}_G$$

$$= \tfrac{1}{2}M[a\omega(0 \quad 0 \quad -1)]^2 + \tfrac{1}{2}\frac{\omega}{\sqrt{2}}(1 \quad -1 \quad 0).\frac{Ma^2\omega}{4\sqrt{2}}(1 \quad -1 \quad 0)$$

$$= \tfrac{5}{8}Ma^2\omega^2.$$

Alternatively using (RB6′) we have

$$T = \tfrac{1}{2}\boldsymbol{\omega}.\mathbf{L}_Q$$

$$= \tfrac{1}{2}\frac{\omega}{\sqrt{2}}(1 \quad -1 \quad 0).\frac{5Ma^2\omega}{4\sqrt{2}}(1 \quad -1 \quad 0)$$

$$= \tfrac{5}{8}Ma^2\omega^2$$

as before.

6.8   Summary

To conclude our investigation of the equations satisfied by a rigid body in its motion we collect together the important results. First consider general motion.

GENERAL MOTION

(RB1)
$$\mathbf{v_P} = \mathbf{v} + \boldsymbol{\omega} \wedge \mathbf{r}$$

where $\mathbf{r} = \underline{\mathrm{GP}}$ and $\mathbf{v}$ is the velocity of G.

(RB2)
$$M\frac{d\mathbf{v}}{dt} = \sum \mathbf{F}_i$$

where $\mathbf{v}$ is the velocity of G.

(RB3)
$$\frac{d\mathbf{L_G}}{dt} = \sum \mathbf{r}_i \wedge \mathbf{F}_i$$

where $\mathbf{r}_i$ is a position vector the line of action of the external force $\mathbf{F}_i$.

(RB4)
$$\mathbf{L_G} = \mathfrak{I}_G \boldsymbol{\omega}$$

where $\mathfrak{I}_G$ is the inertia tensor at G and $\boldsymbol{\omega}$ is the angular velocity of the body.

(RB5)
$$\mathbf{L_P} = \mathbf{s} \wedge M\mathbf{v} + \mathbf{L_G}$$

where $\mathbf{s} = \underline{\mathrm{PG}}$.

(RB6)
$$T = \tfrac{1}{2}M\mathbf{v}^2 + \tfrac{1}{2}\boldsymbol{\omega}.\mathbf{L_G}.$$

This last completes the set of equations which allow us to caluclate the motion of a rigid body and which are a direct extension of Newton's laws for a particle. We can see that the solution of these equations involves finding both the velocity $\mathbf{v}$ of the centre of mass G and also the angular velocity $\boldsymbol{\omega}$ of the body – six scalar velocities to be found in terms of the time t in general. The number of scalars reflects the number of degrees of freedom of a rigid body in unrestricted motion.

ONE POINT ANCHORED

In the case of a body rotating about a point Q of itself which is anchored, rather simpler equations are satisfied, as we have seen:

(RB3′)
$$\dot{\mathbf{L}}_Q = \sum \mathbf{s}_i \wedge \mathbf{F}_i$$

where $\mathbf{s}_i$ is a position vector from Q to the line of action of the external force $\mathbf{F}_i$.

(RB4′)
$$\mathbf{L}_Q = \mathfrak{I}_Q\boldsymbol{\omega}.$$

(RB6′)
$$T = \tfrac{1}{2}\boldsymbol{\omega}.\mathbf{L}_Q.$$

These three equations enable a solution for $\boldsymbol{\omega}$ to be found in many examples of rotation about a fixed point, determining three scalars as functions of t. This is in line with the reduction to three degrees of freedom for a body with one point anchored. The equations of general motion still hold but introduce more complexity than is needed. We may also need to use (6.1) in the altered form below:

(RB1′)
$$\mathbf{v}_P = \boldsymbol{\omega} \wedge \mathbf{p}$$

where $\mathbf{p} = \underline{QP}$.

Exercises:
Chapter 6

1. A rigid body is acted on by a set of forces. Show that if the vector sum of the forces is zero and the forces have no moment about G at any time, then the body will remain in equilibrium if it is stationary at time $t = t_0$.

2. A pair of forces $\mathbf{F}$ and $-\mathbf{F}$ act on a rigid body. They are applied at two points of the body, P, Q respectively, and $\underline{PQ} = \mathbf{b}$. Prove that the moment of this system of forces is the same about any point. Such a system of forces is called a *couple*. Under what condition involving $\mathbf{b}$ and $\mathbf{F}$ could the rigid body be in equilibrium if acted on solely by these two forces?

3. Suppose that a system of forces $\mathbf{F}_i$, $i = 1, 2, \ldots, n$, acts on a rigid body. Each force $\mathbf{F}_i$ acts along a line one of whose points has position vector $\mathbf{r}_i$ measured from a point P. Show that the moment of the system of forces about any other point Q is the same as that of a new system composed of a resultant force $\mathbf{F} = \Sigma\,\mathbf{F}_i$ acting at P and a couple $\boldsymbol{\Gamma} = \Sigma\,\mathbf{r}_i \wedge \mathbf{F}_i$. What are the equivalent force and couple at a point position vector $\mathbf{a}$ from P for the same system of forces? Deduce that if any system of forces and couples acting on a rigid body has zero resultant force and zero moment about some arbitrary point P then the rigid body can remain in equilibrium.

   Four forces act along the inward normals to the faces of a tetrahedron, each line of action passing through the opposite vertex and the magnitude of each force being a multiple k of the area of the respective face. Prove that the tetrahedron can remain in equilibrium.

4. A rigid body is moving with one point O fixed in an inertial frame and fixed in the body. If the body has angular velocity $\boldsymbol{\omega}$ and if P is a point with position vector $\mathbf{r}$ measured from O prove that the acceleration of P, $\dot{\mathbf{v}}_P$, satisfies

$$\dot{\mathbf{v}}_P = \dot{\boldsymbol{\omega}} \wedge \mathbf{r} + \boldsymbol{\omega} \wedge (\boldsymbol{\omega} \wedge \mathbf{r}).$$

   Hence show that if $\boldsymbol{\omega} \wedge \dot{\boldsymbol{\omega}} \neq \mathbf{0}$, then O is the only point of the body which is not accelerating.
   [Hint: we may write any vector $\mathbf{r} = \lambda\boldsymbol{\omega} + \mu\dot{\boldsymbol{\omega}} + \nu\boldsymbol{\omega} \wedge \dot{\boldsymbol{\omega}}$. Find equations for λ, μ, ν in the case $\dot{\mathbf{v}}_P = \mathbf{0}$ and show that their only solution is $\lambda = \mu = \nu = 0$.]

5. Using multiple integrals calculate the inertia tensor at the centre of mass of the following uniform bodies:

(a) solid sphere of radius a and mass m;

(b) solid cylinder, height h, radius a, and mass m;

(c) a flat rectangular lamina, sides $2a$, $2b$, and mass m.

Use symmetry wherever possible to cut down the amount of integration involved.

6. (a) Consider a uniform cube, side $2a$, mass m. Given that the inertia tensor at the centre of mass is given by

$$\Im_G = \tfrac{2}{3}ma^2 I$$

where I is the identity matrix, calculate the inertia tensor at a vertex using the parallel axis theorem.

(b) If the cube is rotating with angular velocity ω, the instantaneous axis of rotation being along an edge of the cube, calculate the angular momentum about a vertex on the axis of rotation.

(c) If the cube is rotating about an axis parallel to an edge and through its centre of mass calculate the angular momentum at a vertex.

(d) In each of cases (b) and (c) find the kinetic energy of the cube.

7. Find the principal moments of inertia and the prinicpal axes at a vertex of a uniform solid cube.

8. Consider a solid uniform cone, mass m, height h, and radius of base a.

(a) Find its centre of mass.

(b) Calculate the inertia tensor at the vertex of the cone.

(c) Hence calculate the inertia tensor at the centre of mass. Under what conditions on h and a does the cone have dynamical symmetry at its centre of mass (meaning that the inertia tensor at the centre of mass is a multiple of the identity matrix)?

9. Prove that if an axis passes through the centre of mass G and is a principal axis at G then it is a principal axis at any point along it also.

At a point Q on the rim of a uniform cone use basic symmetry of the cone to show that the tangent to the rim at Q must be a principal axis. If in addition the cone has dynamical symmetry at its centre of mass (see question 8) find the directions of all the principal axes at Q.

7.1 Bodies with dynamical symmetry

A sphere is symmetrical under any rotation about its centre of mass. A cube does not have the total symmetry of a sphere. Yet if it is rotating about its centre of mass G its behaviour is very similar to that of the sphere, due to the simple form of the inertia tensor at G. In each case the inertia tensor is a multiple of the identity matrix.

Definition 7.1 A rigid body has *dynamical symmetry* about a point P if the inertia tensor at P is a multiple of the identity. It follows that all axes at P are principal axes.

In a case of a rigid body with dynamical symmetry at its centre of mass the angular momentum at G has an extremely simple form

$$\mathbf{L_G} = \mathfrak{I}_G \boldsymbol{\omega}$$

$$= A\boldsymbol{\omega}$$

using (RB4) together with the fact that $\mathfrak{I}_G = AI$ where A is the moment of inertia about any axis through G, and I is the identity matrix. This means that the angular momentum at G is simply a scalar multiple of the angular velocity.

Example 1 A regular tetrahedron is rotating about its centre of mass G which is fixed. If it is rotating under the influence of a frictional couple $-\lambda\boldsymbol{\omega}$, $\lambda > 0$, where $\boldsymbol{\omega}$ is the angular velocity of the body, calculate the angular velocity of the tetrahedron at time t, given that $\boldsymbol{\omega} = \boldsymbol{\omega}_0$ at $t = 0$. Show that the kinetic energy decays exponentially with time.

Take axes as shown in the diagram. We first need to consider the inertia tensor at the centre of mass. Assuming that the tetrahedron has a regular mass distribution also then the density will be constant.

Does it have dynamical symmetry at its centre of mass?

Gz is the axis through the vertex, D, Gy lies in the plane GDA and is perpendicular to Gz, and Gx is the axis perpendicular to both so that Gxyz are a Cartesian set.

Since the tetrahedron is symmetrical when reflected in the plane GDA (or Gyz) then for every point (x, y, z) there is a corresponding point $(-x, y, z)$.

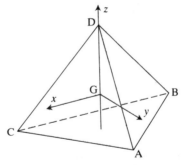

Fig. 7.1

From the definitons

$$H = \sum mxy = 0 = \sum mxz = G$$

using the symmetry. Hence the inertia tensor becomes

$$\Im_G = \begin{bmatrix} A & 0 & 0 \\ 0 & B & -F \\ 0 & -F & C \end{bmatrix}.$$

This matrix has $(1 \quad 0 \quad 0)^T$ as an eigenvector and so Gx must be a principal axis of the body. However, by symmetry there are three such axes, all of them principal with moment of inertia A, none of them at right angles to each other but perpendicular to GD. This gives that any axis perpendicular to GD is a prinicpal axis at G and hence so is GD. Finally D was an arbitrary choice of vertex and hence the inertia tensor is a multiple of the identity

$$\Rightarrow \quad \Im_G = AI$$

where A is the moment of inertia about any axis through G and I is the identity matrix. This result agrees with intuition since the tetrahedron, like the cube, is regular.

To calculate the motion of the body since G is a fixed point we need only calculate the angular velocity.

$$(RB3) \Rightarrow \quad \dot{\mathbf{L}}_G = \sum \mathbf{r} \wedge \mathbf{F} = -\lambda \boldsymbol{\omega}$$

$$(RB4) \Rightarrow \quad \mathbf{L}_G = A\boldsymbol{\omega}.$$

These equations give a very simple differential equation for $\boldsymbol{\omega}$:

$$A \frac{\mathrm{d}}{\mathrm{d}t} \dot{\boldsymbol{\omega}} = -\lambda \boldsymbol{\omega}$$

which on using an integrating factor gives

$$\omega = \omega_0 \exp(-\lambda t/A).$$

Using (RB6) we have that the kinetic energy is given by

$$T = \tfrac{1}{2}\omega.\mathfrak{I}_G\omega$$

so that

$$T = \tfrac{1}{2}A\omega^2 = \tfrac{1}{2}A\omega_0^2 \exp(-2\lambda t/A)$$

and hence the kinetic energy decays exponentially.

7.2   Bodies with axial symmetry: top

Tops are in universal everyday use, e.g. spin drier, lathe, child's humming top. A top is any rigid body with an axis of symmetry. We take first the simplest model, i.e. a top rotating under no forces about its centre of mass.

Example 2   Suppose a top is rotating freely about its centre of mass which is fixed. Show that the instantaneous axis of rotation moves on a cone relative to the inertial frame in which the body rotates and on a cone relative to axes fixed in the body.

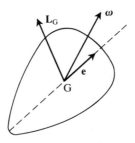

Fig. 7.2

Suppose that e is a unit vector long the axis of symmetry of the top and that ω is the angular velocity. Then the rate of change of the angular momentum at G satisfies (RB3)$\Rightarrow$

$$\frac{d}{dt}L_G = 0.$$

If we use axes fixed in the body with the third axis along $\mathbf{e}$ and any other two perpendicular axes at G then these are principal axes and the inertia tensor at G is

$$\mathfrak{I}_G = \mathrm{diag}\{A, A, C\}.$$

Since $\mathbf{e}$ is fixed in the body it has angular velocity $\boldsymbol{\omega}$ also $\Rightarrow$

$$\dot{\mathbf{e}} = \boldsymbol{\omega} \wedge \mathbf{e}$$

$$\mathbf{e} \wedge \dot{\mathbf{e}} = \mathbf{e} \wedge (\boldsymbol{\omega} \wedge \mathbf{e})$$

$$= \boldsymbol{\omega} - (\boldsymbol{\omega}.\mathbf{e})\mathbf{e}.$$

Define the *spin* about the axis of symmetry to be

$$n = \boldsymbol{\omega}.\mathbf{e}$$

and we have

$$\boldsymbol{\omega} = \mathbf{e} \wedge \dot{\mathbf{e}} + n\mathbf{e}. \tag{7.1}$$

Now the second term on the right lies along the axis of symmetry and is therefore along a principal axis. The first term is perpendicular to the axis of symmetry and also lies along a principal axis since any axis at G perpendicular to the axis of symmetry is a principal axis. Hence we have from (RB4)

$$\mathbf{L}_G = \mathfrak{I}_G \boldsymbol{\omega}$$

$$= A\mathbf{e} \wedge \dot{\mathbf{e}} + Cn\mathbf{e}. \tag{7.2}$$

Given that $\dot{\mathbf{L}}_G = \mathbf{0}$ two results immediately follow:

(i) $\mathbf{L}_G$ is a constant vector in the inertial frame

(ii) $\dot{\mathbf{L}}_G = A\mathbf{e} \wedge \ddot{\mathbf{e}} + C\dot{n}\mathbf{e} + Cn\dot{\mathbf{e}} = \mathbf{0}$. $\tag{7.3}$

Since $\mathbf{e}$ is a unit vector it is perpendicular to $\dot{\mathbf{e}}$. Taking the scalar product of (7.3) with $\mathbf{e}$ gives

$$C\dot{n} = 0 \Rightarrow n \text{ is a } constant.$$

This leaves the equation

$$A\mathbf{e} \wedge \ddot{\mathbf{e}} + Cn\dot{\mathbf{e}} = \mathbf{0}. \tag{7.4}$$

Taking the scalar product with $\mathbf{e} \wedge \dot{\mathbf{e}}$ and integrating gives

$$A(\mathbf{e} \wedge \dot{\mathbf{e}}).(\mathbf{e} \wedge \ddot{\mathbf{e}}) = 0$$

$$\tfrac{1}{2}A(\mathbf{e} \wedge \dot{\mathbf{e}})^2 \text{ is a } constant, \text{ also.}$$

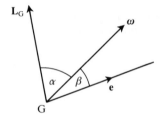

Fig. 7.3

Now using (RB6) for the kinetic energy

$$T = \tfrac{1}{2}\boldsymbol{\omega}.\mathbf{L_G}$$
$$= \tfrac{1}{2}A(\mathbf{e} \wedge \dot{\mathbf{e}})^2 + \tfrac{1}{2}Cn^2$$
$$= \text{constant}. \tag{7.5}$$

To see how the axis of rotation moves consider Fig. 7.3. $\boldsymbol{\omega} = \mathbf{e} \wedge \dot{\mathbf{e}} + n\mathbf{e}$ gives the direction of the axis of rotation, $\mathbf{L_G} = A\mathbf{e} \wedge \dot{\mathbf{e}} + Cn\mathbf{e}$ is a fixed vector in the inertial frame, $\mathbf{e}$ is a vector along the axis of symmetry, and

$$\mathbf{L_G} = A(\boldsymbol{\omega} - n\mathbf{e}) + Cn\mathbf{e} \tag{7.6}$$

$\Rightarrow \mathbf{L_G}, \boldsymbol{\omega}, \mathbf{e}$ are coplanar vectors for all times t.
  Furthermore

$$\cos \alpha = \frac{\boldsymbol{\omega}.\mathbf{L_G}}{\omega L_G} = \frac{2T}{\omega L_G} = \text{constant}$$

so that $\boldsymbol{\omega}$ moves on a cone around $\mathbf{L_G}$ which is fixed in the inertial frame. Relative to axes fixed in the body

$$\cos \beta = \frac{\boldsymbol{\omega}.\mathbf{e}}{\omega} = \frac{n}{\omega} = \text{constant}$$

and $\boldsymbol{\omega}$ moves at a fixed angle to $\mathbf{e}$, the axis of symmetry, and hence also on a cone relative to axes fixed in the body.

Example 3 Suppose that in the above motion we decide to model friction at the pivot G by a couple $-\lambda\boldsymbol{\omega}$, $\lambda > 0$. Prove that in that case the kinetic energy tends to 0 as $t \to \infty$.

Equation (RB3)$\Rightarrow$

$$\dot{\mathbf{L}}_G = -\lambda\boldsymbol{\omega}.$$

Equation (7.3) becomes

$$A\mathbf{e} \wedge \ddot{\mathbf{e}} + C\dot{n}\mathbf{e} + Cn\dot{\mathbf{e}} = -\lambda(\mathbf{e} \wedge \dot{\mathbf{e}} + n\mathbf{e}). \tag{7.7}$$

Scalar product with $\mathbf{e} \Rightarrow$

$$C\dot{n} = -\lambda n \Rightarrow n = n_0 \exp(-\lambda t/C), \ n_0 \text{ constant.}$$

Scalar product with $\mathbf{e} \wedge \dot{\mathbf{e}} \Rightarrow$

$$A(\mathbf{e} \wedge \ddot{\mathbf{e}}).(\mathbf{e} \wedge \dot{\mathbf{e}}) = -\lambda(\mathbf{e} \wedge \dot{\mathbf{e}})^2.$$

If $N = (\mathbf{e} \wedge \dot{\mathbf{e}})^2$ then this equation gives

$$\tfrac{1}{2}A\dot{N} = -\lambda N \Rightarrow N = N_0 \exp(-2\lambda t/A), \ N_0 > 0.$$

But the kinetic energy is given by

$$\begin{aligned} T &= \tfrac{1}{2}A(\mathbf{e} \wedge \dot{\mathbf{e}})^2 + \tfrac{1}{2}Cn^2 \\ &= \tfrac{1}{2}AN_0 \exp(-2\lambda t/A) + \tfrac{1}{2}Cn_0^2 \exp(-2\lambda t/C) \end{aligned} \tag{7.8}$$

and hence the kinetic energy decays exponentially with time.

7.3 Precession

We can now go on to consider the most important application of the motion of a top. Consider a top rotating about a vertex Q on its axis of symmetry. We assume that it is freely pivoted at Q (there will be a reaction at Q but no frictional couple) and that $\mathbf{v}_Q = \mathbf{0}$ throughout the motion. Suppose that the centre of mass is at a distance h along the axis of symmetry from Q. We take $\mathbf{k}$ to be a unit vector pointing vertically upwards. The inertia tensor at Q must be diagonal if we use the axis of symmetry and any two perpendicular axes in the body. Suppose that the principal moments of inertia at Q are A, A, C and that $\mathbf{e}$ is again a unit vector along the axis of symmetry.

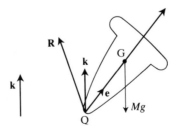

Fig. 7.4

Then applying (RB3′) at Q we have

$$\frac{d\mathbf{L_Q}}{dt} = h\mathbf{e} \wedge (-Mg\mathbf{k}) \qquad (7.9)$$

and as before

$$\boldsymbol{\omega} = \mathbf{e} \wedge \dot{\mathbf{e}} + n\mathbf{e}$$

$$\mathbf{L_Q} = A\mathbf{e} \wedge \dot{\mathbf{e}} + Cn\mathbf{e}$$

giving

$$A\mathbf{e} \wedge \ddot{\mathbf{e}} + C\dot{n}\mathbf{e} + Cn\dot{\mathbf{e}} = -Mgh\mathbf{e} \wedge \mathbf{k}. \qquad (7.10)$$

Again taking the scalar product with $\mathbf{e}$ gives that the spin n is constant

$$n = \text{constant} \qquad (7.11)$$

Both physically and mathematically it is clear that if the top is inclined at an angle to the vertical as shown in Fig. 7.4, it would fall over if it were not spinning. The possible stable patterns of motion are of considerable interest since a spinning top can provide great power, for example drilling through extremely tough materials.

Definition 7.2   Precession occurs if the angle between the axis of symmtery and the vertical is constant.

Definition 7.3   Steady precession occurs if, in addition to a constant inclination to the vertical, the angular velocity of the axis of symmtery about the vertical is constant.

Proposition 7.1   *For a top moving about one point on its axis of symmetry which is fixed, if precession occurs then steady precession occurs.*

Proof   Assume that precession is occurring. Suppose the angle between the axis of symmetry and the vertical is α, a constant.

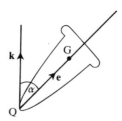

Fig. 7.5

Suppose also that in the *inertial frame* the angular velocity of the axis of symmtery about the vertical is $\Omega\mathbf{k}$ where Ω is not necessarily constant. Then it must be the case that

$$\dot{\mathbf{e}} = \Omega\mathbf{k} \wedge \mathbf{e}.$$

Equation (7.10) gives

$$A\mathbf{e} \wedge \ddot{\mathbf{e}} + Cn\dot{\mathbf{e}} = -Mgh\mathbf{e} \wedge \mathbf{k}. \qquad (7.12)$$

On substituting for $\dot{\mathbf{e}}$

$$\ddot{\mathbf{e}} = \dot{\Omega}\mathbf{k} \wedge \mathbf{e} + \Omega\mathbf{k} \wedge \dot{\mathbf{e}}$$
$$\ddot{\mathbf{e}} = \dot{\Omega}\mathbf{k} \wedge \mathbf{e} + \Omega\mathbf{k} \wedge (\Omega\mathbf{k} \wedge \mathbf{e})$$
$$= \dot{\Omega}\mathbf{k} \wedge \mathbf{e} + \Omega^2 \cos\alpha\mathbf{k} - \Omega^2\mathbf{e}.$$

In (7.12)

$$A\mathbf{e} \wedge \dot{\Omega}(\mathbf{k} \wedge \mathbf{e}) + A\Omega^2 \cos\alpha\mathbf{e} \wedge \mathbf{k} + Cn\Omega\mathbf{k} \wedge \mathbf{e} = -Mgh\mathbf{e} \wedge \mathbf{k} \qquad (7.13)$$

With the exception of the first, every term in (7.13) is along $\mathbf{e} \wedge \mathbf{k}$. The first term is perpendicular to this vector. Hence its coefficient must be zero

$$A\dot{\Omega} = 0$$
$$\Rightarrow \quad \Omega = \text{constant}.$$

This proves the proposition.

Proposition 7.2   *Steady precession at an angle $\alpha \neq 0$ to the vertical with angular velocity Ω about the vertical and spin n about the axis of symmetry can occur if and only if*

$$A\Omega^2 \cos\alpha - Cn\Omega + Mgh = 0.$$

Proof   In (7.13) using the fact that $\dot{\Omega} = 0$ we have

$$(A\Omega^2 \cos\alpha - Cn\Omega + Mgh)\mathbf{e} \wedge \mathbf{k} = 0.$$

Since $\alpha \neq 0$ then $\mathbf{e} \wedge \mathbf{k} \neq \mathbf{0}$ and the result follows.

Corollary   *A solution for Ω exists if and only if*

$$C^2n^2 \geq 4AMgh \cos\alpha.$$

This last result gives a minimum value for n, the spin of the top about its own axis, ensuring that at the fixed angle α there is an angular velocity Ω about the vertical such that the top precesses steadily and does not topple over.

Example 4   Consider a top which is spinning about its vertex Q which is in contact with rough ground. The top is performing steady precession at an angle α to the vertical with angular velocity Ω about it. Calculate the reaction and show that if the ceofficient of static friction is μ then slipping will not occur if a certain inequality is satisfied.

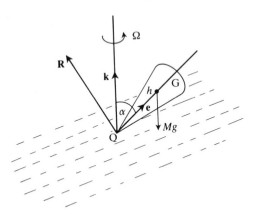

Fig. 7.6

The only forces acting on the top are the force due to gravity and the reaction at Q. Since the reaction has no moment about Q the simplest equation to use is (RB2), the equation for linear momentum at the centre of mass:

$$M\frac{dv}{dt} = \mathbf{R} - Mg\mathbf{k}.$$

Since $\underline{QG} = h\mathbf{e} \Rightarrow \mathbf{v} = h\dot{\mathbf{e}} \Rightarrow \dot{\mathbf{v}} = h\ddot{\mathbf{e}} = h(\Omega^2 \cos\alpha\mathbf{k} - \Omega^2\mathbf{e})$, hence

$$\mathbf{R} = Mg\mathbf{k} + Mh\Omega^2(\cos\alpha\mathbf{k} - \mathbf{e}).$$

The vector $(\cos\alpha\mathbf{k} - \mathbf{e})$ is horizontal and so the normal reaction is Mg, whereas the frictional force is

$$Mh\Omega^2 \sin\alpha.$$

The top can precess without slipping if

$$Mh\Omega^2 \sin\alpha \leq \mu Mg.$$

Example 5   A gyrocompass is composed of a top which is set spinning in a plane which is as nearly as possible horizontal on the surface of the Earth. It is held in such a way that its axis of symmetry can only move in this horizontal plane and its centre of mass is stationary relative to the Earth's surface. We look for the positions in which it spins about its own axis

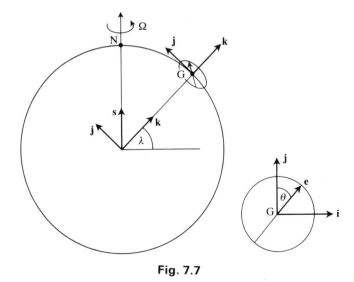

Fig. 7.7

with that axis stationary relative to the Earth's surface. We are looking
for positions of relative equilibrium. Essentially the compass uses the
spin of the Earth itself which permits relative equilibrium in specific
directions only. Relative equilibrium means stationary in the frame fixed
relative to the Earth.

Take axes $\mathbf{i}$, $\mathbf{j}$, $\mathbf{k}$ fixed relative to the surface of the Earth where $\mathbf{k}$ is
vertically upwards from the surface, $\mathbf{j}$ points horizontally due north
leaving $\mathbf{i}$ along the tangent to the circle of latitude on the Earth's surface.
Suppose that the compass is at latitude λ and that the axis of symmetry
is inclined at an angle θ to $\mathbf{j}$ as shown in Fig. 7.7. Assume $\mathbf{e}$ to be a unit
vector along the axis of symmtery and n to be the spin of the top about
its own axis. The axes $\mathbf{i}$, $\mathbf{j}$, $\mathbf{k}$ are fixed in the Earth and are spinning with
it. The inertial frame at the centre of the Earth can be thought of as fixed
relative to the Sun. This is a good approximation to an inertial frame given
the time scale over which the compass is observed. The forces are arranged
so that they have components only in the directions of $\mathbf{e}$ and $\mathbf{k}$. Then with
respect to the rotating frame

$$\boldsymbol{\omega} = -\dot{\theta}\mathbf{k} + n\mathbf{e} \qquad \mathbf{L}_G = -A\dot{\theta}\mathbf{k} + Cn\mathbf{e}.$$

Equation (RB3)$\Rightarrow$

$$\frac{\mathrm{d}\mathbf{L}_G}{\mathrm{d}t} = a\mathbf{e} \wedge (\mathbf{I} - \mathbf{J})$$

where $\mathbf{I}$, $\mathbf{J}$ are forces parallel to $\mathbf{k}$, ensuring the compass is horizontal.

The axes $\mathbf{i}, \mathbf{j}, \mathbf{k}$ are rotating with the angular velocity of the Earth, which is $\Omega \mathbf{s}$ as shown in the diagram. Hence by the rotating axes theorem

$$\frac{d\mathbf{L}_G}{dt} = \left(\frac{d\mathbf{L}_G}{dt}\right)' + \Omega \mathbf{s} \wedge \mathbf{L}_G.$$

This gives

$$-A\ddot{\theta}\mathbf{k} + C\dot{n}\mathbf{e} + Cn\dot{\mathbf{e}} + \Omega \mathbf{s} \wedge (-A\dot{\theta}\mathbf{k} + Cn\mathbf{e}) = a\mathbf{e} \wedge (\mathbf{I} - \mathbf{J}). \tag{7.14}$$

Looking at the component in the direction $\mathbf{e} \Rightarrow$

$$C\dot{n} - A\Omega\dot{\theta}\mathbf{s} \wedge \mathbf{k}.\mathbf{e} = 0$$

or

$$C\dot{n} - A\Omega\dot{\theta}\cos\lambda\sin\theta = 0. \tag{7.15}$$

In equilibrium relative to the Earth's surface $\theta = $ constant. Hence $n = $ constant.

Taking the scalar product of (7.14) with respect to $\mathbf{k}$ gives

$$-A\ddot{\theta} + Cn\Omega\mathbf{s} \wedge \mathbf{e}.\mathbf{k} = 0. \tag{7.16}$$

Since $\ddot{\theta} = 0$ in relative equilibrium and since n, Ω are non-zero $\Rightarrow \mathbf{s}, \mathbf{e}, \mathbf{k}$ are coplanar. Since $\mathbf{e}$ is horizontal this indicates that it must lie along the north–south line, $\mathbf{j}$, and we have $\theta = 0$ or π, so that the compass points along the north–south line. NB There are two points on the Earth's surface for which this is not the case! Which are they?

Given that a ship using a gyrocompass is unlikely to be stationary it is essential that the positions of relative equilibrium are stable. Given that $n > 0$ consider the position $\theta = 0$. Let θ be small and assume that n is constant (a reasonable approximation from (7.15)).

In (7.16) we have

$$\mathbf{s} \wedge \mathbf{e}.\mathbf{k} = -\mathbf{s} \wedge \mathbf{k}.\mathbf{e}$$

$$= -\cos\lambda\sin\theta$$

and

$$-A\ddot{\theta} - Cn\Omega\sin\theta\cos\lambda = 0.$$

For θ small

$$\ddot{\theta} + \left(\frac{Cn\Omega}{A}\cos\lambda\right)\theta = 0$$

indicating stable equilibrium, and since clearly equilibrium for $\theta = (\pi + \varepsilon)$ is unstable this indicates that the compass points due north in practical use.

We will consider one further example in this section on tops.

Example 6   A top, whose principal moments of inertia at its vertex P are A, A, C, rotates about its vertex, which is forced to move so that at time t its position vector is $\mathbf{a}(t)$ with respect to a fixed origin O. The top is of mass M and the distance of its centre of mass from P is h.

Obtain equations of motion of the top in terms of a unit vector along its axis of symmtery.

The top is rotating about its vertex on the perfectly rough floor of a lift, which is accelerating downwards with acceleration λ, a constant. Show that the top can precess about the vertical with a constant angular speed Ω, at an angle α where

$$A\Omega^2 \cos \alpha - Cn\Omega + Mh(g - \lambda) = 0.$$

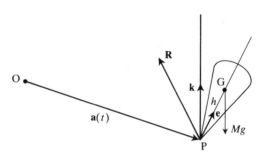

Fig. 7.8

Since in this example P is not a fixed point we must use the general equations at G. Set up the geometry as before with $\mathbf{e}$ a unit vector along the axis of symmtery and n the spin about this axis. Then

$$\mathbf{v_G} = h\dot{\mathbf{e}} + \dot{\mathbf{a}} \quad \text{since } \underline{\text{OG}} = h\mathbf{e} + \mathbf{a}$$

$$\boldsymbol{\omega} = \mathbf{e} \wedge \dot{\mathbf{e}} + n\mathbf{e}.$$

Equation (RB2) gives

$$M\frac{\mathrm{d}\mathbf{v_G}}{\mathrm{d}t} = \mathbf{R} - Mg\mathbf{k} \tag{7.17}$$

and from (RB3)

$$\frac{\mathrm{d}\mathbf{L_G}}{\mathrm{d}t} = -h\mathbf{e} \wedge \mathbf{R} \tag{7.18}$$

where $\mathbf{R}$ is the reaction at P.

The moment of inertia are given at the vertex P so that in order to write down the equation $\mathbf{L}_G = \mathfrak{I}_G \boldsymbol{\omega}$ we need to employ the parallel axis theorem to calculate $\mathfrak{I}_G$. Using $\mathfrak{I}_P = \mathfrak{I}_G + M(\mathbf{p}^T\mathbf{p}I - \mathbf{p}\mathbf{p}^T)$ gives

$$A' = A - Mh^2 \qquad C' = C$$

for the principal moments of inertia at G.

Hence as before

$$\mathbf{L}_G = (A - Mh^2)\mathbf{e} \wedge \dot{\mathbf{e}} + Cn\mathbf{e}.$$

Putting this expression for the angular momentum into (7.18)

$$(A - Mh^2)\mathbf{e} \wedge \ddot{\mathbf{e}} + Cn\dot{\mathbf{e}} + C\dot{n}\mathbf{e} = -h\mathbf{e} \wedge \mathbf{R}$$

and substituting for $\mathbf{R}$ from (7.17) gives

$$(A - Mh^2)\mathbf{e} \wedge \ddot{\mathbf{e}} + Cn\dot{\mathbf{e}} + C\dot{n}\mathbf{e} = -Mh\mathbf{e} \wedge (h\ddot{\mathbf{e}} + \ddot{\mathbf{a}} + g\mathbf{k}). \qquad (7.19)$$

Taking the scalar product with $\mathbf{e}$ tells us that $n = $ constant, again, and cancelling terms we have

$$A\mathbf{e} \wedge \ddot{\mathbf{e}} + Cn\dot{\mathbf{e}} = -Mh\mathbf{e} \wedge \ddot{\mathbf{a}} - Mgh\mathbf{e} \wedge \mathbf{k}. \qquad (7.20)$$

If the top is precessing then as before

$$\mathbf{e}.\mathbf{k} = \cos\alpha \quad \text{and} \quad \dot{\mathbf{e}} = \Omega\mathbf{k} \wedge \mathbf{e}.$$

Also in the lift

$$\ddot{\mathbf{a}} = -\lambda\mathbf{k}.$$

Remembering that

$$\mathbf{e} \wedge \ddot{\mathbf{e}} = \Omega^2 \cos\alpha\, \mathbf{e} \wedge \mathbf{k}$$

$(7.20) \Rightarrow$

$$A\Omega^2 \cos\alpha - Cn\Omega + Mh(g - \lambda) = 0.$$

It is exciting to see that this equation agrees precisely with expectations on physical grounds since the effect of the downward acceleration of the lift is to reduce the effective acceleration due to gravity to $g - \lambda$.

7.4   General rotation of a rigid body fixed at one point

To extend the scope of this chapter we can consider the rotational motion of a body with no special symmetry properties about a single pivot point P. Since P is a fixed point of the body it is clear that (RB3′) applies:

$$\frac{dL_P}{dt} = \sum \mathbf{r} \wedge \mathbf{F} = \Gamma \tag{7.21}$$

where Γ is the total moment of all forces (and couples) about P.

We can use (RB4′) so that $L_P = \Im_P\omega$, but the resulting equations will be difficult to deal with for a general set of axes. *Euler's equations* follow if the axes chosen are the principal axes of the body at P. These axes are helpful since the inertia tensor is diagonal with respect to them. We must use the rotating axes theroem as the principal axes rotate in an inertial frame with the same angular velocity as the body. If $\partial/\partial t$ denotes rate of change with respect to the axes fixed in the body then (7.21) becomes

$$\frac{\partial L_P}{\partial t} + \omega \wedge L_P = \Gamma. \tag{7.22}$$

Writing $\omega = (\omega_1, \omega_2, \omega_3)$ along the principal axes then eqn (RB4′) tells us that

$$L_P = (A\omega_1, B\omega_2, C\omega_3)$$

where A, B, C are the principal moments of inertia at P. If $\Gamma = (\Gamma_1, \Gamma_2, \Gamma_3)$ with respect to the same axes then (7.22) results in the following scalar equations:

$$A\dot{\omega}_1 - (B - C)\omega_2\omega_3 = \Gamma_1.$$
$$B\dot{\omega}_2 - (C - A)\omega_3\omega_1 = \Gamma_2$$
$$C\dot{\omega}_3 - (A - B)\omega_1\omega_2 = \Gamma_3. \tag{7.23}$$

These are *Euler's equations* for the rotational motion of a body about one point of itself which is pivoted.

Example   *Special case of body rotating freely about its centre of mass.* If the pivot is situated at the centre of mass and no external forces other than gravity are applied then $\Gamma = 0$ provided there is no frictional couple at the pivot. The techincal phrase describing absence of a frictional couple is 'freely pivoted'. Of course in practice this is virtually impossible to achieve unless considering space travel! Nevertheless we can learn a good deal about the physical properties of a rotating system from the equations for a freely pivoted body. A possible application would be to a space station spinning about an axis through its centre of mass. If $\Gamma = 0$ then eqn. (7.23) become

$$A\dot{\omega}_1 - (B - C)\omega_2\omega_3 = 0$$
$$B\dot{\omega}_2 - (C - A)\omega_3\omega_1 = 0$$
$$C\dot{\omega}_3 - (A - B)\omega_1\omega_2 = 0. \tag{7.24}$$

We will assume without loss of generality that $A > B > C$ and consider the general case of a body with no inertial symmetry. Then these equations have three steady state solutions, corresponding to solutions for which $\dot{\boldsymbol{\omega}} = \mathbf{0}$. These solutions are $\omega_1 =$ constant, $\omega_2 = 0$, $\omega_3 = 0$, and similarly with ω_2 or ω_3 non-zero, the other two components being zero. The steady state solutions are those solutions in which the body spins with constant angular velocity about a principal axis. It would be useful to know which of these solutions is stable. It would not be sensible to set a space station spinning about an unstable axis!

STABILITY

Suppose that the steady state corresponds to $\omega_1 = n$, a constant, with ω_2 and ω_3 both zero. Then consider a small perturbation from this state, which is to say

$$\omega_1 = n + \varepsilon_1 \qquad \omega_2 = \varepsilon_2 \qquad \omega_3 = \varepsilon_3.$$

Equations (7.24) become

$$A\dot{\varepsilon}_1 - (B - C)\varepsilon_2\varepsilon_3 = 0$$
$$B\dot{\varepsilon}_2 - (C - A)(n + \varepsilon_1)\varepsilon_3 = 0$$
$$C\dot{\varepsilon}_3 - (A - B)(n + \varepsilon_1)\varepsilon_2 = 0. \qquad (7.25)$$

Correct to first order in the small perturbations ε_1 can be taken to be zero and the two remaining equations become

$$B\dot{\varepsilon}_2 - (C - A)n\varepsilon_2 = 0$$
$$C\dot{\varepsilon}_3 - (A - B)n\varepsilon_2 = 0.$$

If we eliminate ε_3 by differentiation and substitution then

$$BC\ddot{\varepsilon}_2 - (C - A)(A - B)n^2\varepsilon_2 = 0.$$

As $A > B > C$ we have

$$\ddot{\varepsilon}_2 + \frac{(A - C)(A - B)n^2}{BC}\varepsilon_2 = 0$$

giving simple harmonic motion since the coefficient of ε_2 is positive. The equation satisfied by ε_3 is identical. These approximate solutions tell us that steady rotation about the principal axis with largest moment of inertia is *stable*.

Considering motion about the middle axis, moment of inertia B, then ε_1 satisfies

$$\ddot{\varepsilon}_1 + \frac{(B - C)(B - A)n^2}{CA}\varepsilon_1 = 0$$

and in this case the coefficient is negative, ensuring that this motion is *unstable*.

Similarly we can conclude that the motion about the principal axis with smallest moment is *stable*.

Applying these results to a uniform cuboid, spinning about an axis parallel to either the longest or the shortest edge, through the centre of mass, gives a stable motion, whereas spinning about an axis parallel to the edge of middle length is an unstable motion of the body.

We can draw one further conclusion from these equations. The rotational energy for the general free motion about the centre of mass is

$$T_R = \tfrac{1}{2}\boldsymbol{\omega}.\mathbf{L}_G$$

$$= \tfrac{1}{2}(A\omega_1^2 + B\omega_2^2 + C\omega_3^2).$$

Differentiating we have

$$\frac{\mathrm{d}T_R}{\mathrm{d}t} = A\dot{\omega}_1\omega_1 + B\dot{\omega}_2\omega_2 + C\dot{\omega}_3\omega_3$$

$$= 0 \qquad\qquad (7.26)$$

on substituting from eqns (7.24). This states that the rotational energy is conserved in a general motion with no moment of forces about **G**.

One other scalar function is conserved in the general motion and that is $\mathbf{L}_G^2$:

$$\mathbf{L}_G^2 = \mathbf{L}_G.\mathbf{L}_G$$

$$= A^2\omega_1^2 + B^2\omega_2^2 + C\omega_3^2.$$

Again differentiating we have

$$\frac{\mathrm{d}(\mathbf{L}_G^2)}{\mathrm{d}t} = A^2\dot{\omega}_1\omega_1 + B^2\dot{\omega}_2\omega_2 + C_j^2\dot{\omega}_3\omega_3 \qquad (7.27)$$

$$= 0$$

again on substituting from eqns (7.24).

These two equations provide us with two first integrals of Euler's equations which can be used to eliminate, say, ω_2 and ω_3 from (7.24) leaving an equation in ω_1 which in principle can be solved analytically. If in addition we have some initial data then the values of the constants T_R and $\mathbf{L}_G^2$ will be known, leading to a precise solution for ω as a function of time.

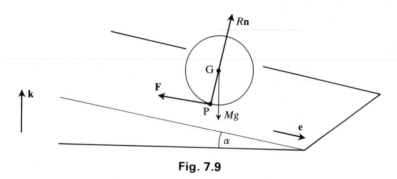

Fig. 7.9

7.5   Rolling spheres

We will finish this chapter by considering briefly the motion of a rolling sphere, possibly a snooker ball rolling on a table which has been incorrectly levelled! By rolling we shall mean that the point of contact of the sphere with the table has the same instantaneous velocity regarded as a point of the table or as a point of the sphere. A simpler way of saying this is that the point of contact of the sphere is instantaneously at zero velocity, since the table is not in motion. Note that the point of contact changes all the time. In Fig. 7.9 α is the inclination of the plane to the horizontal, $\mathbf{n}$ is a unit vector perpendicular to the plane, $\mathbf{e}$ a unit vector down the line of greatest slope, and $\mathbf{k}$ is a unit vector directed vertically upwards. We assume that the sphere is uniform so that the centre of mass G is at the centre of the sphere. There are only two forces acting on the sphere: gravity and the reaction at the point of contact. In the diagram the latter force is split into two components: a normal reaction $R\mathbf{n}$ and a frictional force which is parallel to the plane but whose direction is otherwise unknown. The motion of the sphere can be described by two vectors, $\mathbf{v}$ the velocity of the centre of mass and $\boldsymbol{\omega}$ the angular velocity of the sphere. The equations which we use are (RB2) linear momentum at G, and (RB3), angular momentum about G. The sphere has a single moment of inertia about any axis through the centre of mass. We know that its value is $\frac{2}{5}Ma^2$ where M is the mass and a is the radius. In the interests of simplicity let this moment of inertia be A.

(RB2) gives

$$M\frac{d\mathbf{v}}{dt} = -Mg\mathbf{k} + R\mathbf{n} + \mathbf{F} \tag{7.28}$$

and (RB3)

$$A\frac{d\mathbf{w}}{dt} = -a\mathbf{n} \wedge \mathbf{F} \tag{7.29}$$

since $\mathbf{L_G} = A\omega$.

At this stage we have not used the rolling condition. To do this note that if $\mathbf{v_P}$ is the velocity of the point of contact (the point on the surface of the sphere which is instantaneously in contact with the plane) then $\mathbf{v_P} = \mathbf{0}$. We have an equation (RB1) which tells us how to find this velocity in terms of $\mathbf{v}$ and ω:

$$\mathbf{v_P} = \mathbf{v} + \omega \wedge (-a\mathbf{n}) = \mathbf{0}. \tag{7.30}$$

If we make the sensible assumption that the sphere does not jump off the plane but maintains contact, then both $\mathbf{v}$ and its derivative are parallel to the plane, although note that they need not be directed along the line of greatest slope. Hence considering the components normal to the plane in (7.28) gives

$$-Mg\mathbf{k}.\mathbf{n} + R = 0$$

or

$$R = Mg\cos\alpha$$

and hence we have

$$M\frac{d\mathbf{v}}{dt} = Mg\sin\alpha\mathbf{e} + \mathbf{F} \tag{7.31}$$

on writing $\mathbf{k} = \cos\alpha\mathbf{n} - \sin\alpha\mathbf{e}$.

Differentiating (7.30)

$$\mathbf{0} = \frac{d\mathbf{v}}{dt} - \frac{d\mathbf{w}}{dt} \wedge a\mathbf{n}$$

$$\mathbf{0} = g\sin\alpha\mathbf{e} + M^{-1}\mathbf{F} + A^{-1}(a\mathbf{n} \wedge \mathbf{F}) \wedge a\mathbf{n}.$$

Since $\mathbf{F}$ is perpendicular to $\mathbf{n}$ the last product is simply $A^{-1}a^2\mathbf{F}$ and so using the known value of A

$$\mathbf{F} = -\tfrac{5}{7}Mg\sin\alpha\mathbf{e}. \tag{7.32}$$

If the surface has a ceofficient of static friction μ this motion can only take place if $|\mathbf{F}| \le \mu R$, giving the condition

$$\tfrac{5}{7}\tan\alpha \le \mu.$$

Substituting for $\mathbf{F}$ in eqn (7.31) for $\mathbf{v}$ gives

$$M\frac{d\mathbf{v}}{dt} = Mg\sin\alpha\mathbf{e} - \tfrac{5}{7}Mg\sin\alpha\mathbf{e}$$

$$= \tfrac{2}{7}Mg\sin\alpha\mathbf{e}. \tag{7.33}$$

Suppose that the sphere is set rolling from an origin which initially coincides with G and that the velocity of the centre of mass is $V\mathbf{j}$ where $\mathbf{j}$ is a unit vector both horizontal and parallel to the inclined plane. This means that the sphere is projected horizontally across the plane. What does the path of the sphere look like? We can take $\mathbf{r}$ to be the position vector of G at time t and choose $\mathbf{r} = \mathbf{0}$ at $t = 0$. The initial conditions are: at $t = 0$, $\mathbf{r} = \mathbf{0}$ and $\mathbf{v} = \dot{\mathbf{r}} = V\mathbf{j}$. Integrating (7.33) twice we have

$$\mathbf{r} = \tfrac{1}{7}gt^2 \sin\alpha\,\mathbf{e} + Vt\mathbf{j}. \tag{7.34}$$

Since the component down the plane is a function of t^2 and the component across the plane is a function of t only, this shows that the path of the centre of mass is a parabola whose plane is parallel to the inclined plane.

This example is just one of many concerning rigid bodies rolling. The main points to note are (i) that a body's motion can be completely described if we first solve for $\mathbf{v}$ and $\boldsymbol{\omega}$, and (ii) that two bodies are in rolling contact if the points on each which are instantaneously in contact have the same velocity at the instant of contact.

Exercises:
Chapter 7

1. A body is rotating about its centre of mass with angular velocity $\boldsymbol{\omega}$, under the influence of a constant couple $\mathbf{G}_0$ and a variable couple $-\lambda\boldsymbol{\omega}$, where λ is a positive constant. The inertia tensor at G is a multiple of the identity $\mathfrak{I}_G = AI$. If at time $t = 0$, $\omega = \omega_0$, derive $\boldsymbol{\omega}$ at time t and deduce its limiting direction as $t \to \infty$.

2. An axisymmetric body is rotating freely about its centre of mass G which is fixed. Assume the principal moments of inertia at G are A, A, C.
   Prove that
   (a) the kinetic energy is constant,
   (b) the axis of symmetry moves on a cone in space,
   (c) if the magnitude of the angular momentum about G is L, prove that the time taken by the axis of symmetry to traverse this cone once is $2\pi A/L$. [To tackle this part, notice that the angular velocity of the axis of symmetry about $\mathbf{L}_G$ can also be written as $\Omega\hat{\mathbf{L}}_G$ relative to the inertial frame. Hence find Ω.]

3. Show that, when a rigid body is pivoted at a fixed point O, its kinetic energy can be expressed as $\tfrac{1}{2}\boldsymbol{\omega}\cdot\mathbf{L}_O$ where $\boldsymbol{\omega}$ is the angular velocity of the body and $\mathbf{L}_O$ is the angular momentum about O.

   A top is freely pivoted at a point O on its axis of summetry. The mass of the top is M, its principal moments of inertia at O are A, A, C, and the distance of its centre of mass from O is h. The top is precessing steadily about the vertical with angular speed of precession Ω, n being the spin of the top about its axis of symmetry. Show that the axis of the body is inclined at an angle

$$\cos^{-1}\left(\frac{Cn\Omega - Mgh}{A\Omega^2}\right)$$

to the upward vertical and that the kinetic energy of the top is

$$\tfrac{1}{2}(A\Omega^2)^{-1}[A\Omega^2(Cn^2 + A\Omega^2) - (Cn\Omega - Mgh)^2]$$

(Oxford)

4. A top rotates under the influence of a couple $-\lambda n\mathbf{e}$ about G, the centre of mass, $\lambda > 0$. Show that the spin n about its axis of symmetry satisfies

$$n = \alpha\exp(-\lambda t/C)$$

where α is a constant, C is the moment of inertia about the axis of symmetry, and $\mathbf{e}$ is a unit vector along the axis. Hence prove that the kinetic energy has the form

$$\tfrac{1}{2}C\alpha^2\exp(-2\lambda t/C) + \text{constant}.$$

5. A symmetrical top of mass M is freely pivoted at a point O on its axis of symmetry. The principal moments of inertia at O are A, A, C, and distance of the centre of mass G from O is h. Let $\mathbf{k}$ be a unit vector in the direction of the upward vertical, let $\mathbf{e}$ be a unit vector in the direction of OG, and let n be the component along $\mathbf{e}$ of the angular velocity $\boldsymbol{\omega}$ of the body. If $\mathbf{k} \wedge \mathbf{e} \neq \mathbf{0}$ then show that the following quantities are constant throughout the motion:

$$n; \qquad Cn\mathbf{k}.\mathbf{e} + A\mathbf{k}.(\mathbf{e} \wedge \dot{\mathbf{e}}); \qquad A(\mathbf{e} \wedge \dot{\mathbf{e}})^2 + 2Mgh\mathbf{k}.\mathbf{e}.$$

(Oxford)

6. Write down Euler's equations for a top, principal moments of inertia A, A, C at G, and rotating under no forces about G, its centre of mass. Prove that $\omega_3 = n$, a constant, and that both the other components of angular velocity perform simple harmonic motion with period $2\pi A/(n|C - A|)$.

7. A 10p piece is rotating in contact with a shiny bar top. Model the coin as a disc with radius a and principal moments of inertia A, A, C at the centre of mass G. If the contact between the bar top and the 10p piece is regarded as approximately smooth write down the equations for rate of change of linear momentum and angular momentum at G. Deduce that if the coin is set in motion in such a way that the centre of mass G has no horizontal component of velocity then G can only move up and down a fixed vertical line. Show that

$$A\mathbf{e} \wedge \ddot{\mathbf{e}} + Cn\dot{\mathbf{e}} + C\dot{n}\mathbf{e} = -a\mathbf{s} \wedge \left(M\frac{d\mathbf{v}}{dt} + Mg\mathbf{k}\right) \qquad (*)$$

where $\mathbf{e}$ is a unit normal to the disc, n is the spin of the disc about $\mathbf{e}$, $\mathbf{s}$ is a unit vector in the upward direction of the line of greatest slope in the disc, $\mathbf{v}$ is the velocity of the centre of mass, and $\mathbf{k}$ is a unit vector in the direction of the upward vertical. Hence prove that n is constant.

Show that steady precession, with $\mathbf{e}$ at an angle α to the vertical and angular speed Ω about the vertical, can occur if

$$A\Omega^2 \sin\alpha\cos\alpha - Cn\Omega\sin\alpha - Mga\cos\alpha = 0$$

for $0 < \alpha < \tfrac{1}{2}\pi$. Deduce that for each pair of values (α, n), steady precession is always possible. What is the path traced out on the bar by the point of contact during steady precession?

8. In the previous question use the same model expect for the assumption about contact between the bar and the coin. Assume a very rough contact instead. Deduce that

$$\mathbf{v} + (\mathbf{e} \wedge \dot{\mathbf{e}} + n\mathbf{e}) \wedge (-a\mathbf{s}) = \mathbf{0}.$$

Show that $(*)$ still holds but that $\dot{\mathbf{v}}$ is no longer necessarily vertical. Is it possible that steady precession can occur with the normal to the disc at a fixed angle to the vertical and rotating round the vertical with constant angular speed?

9. Show that if a snooker ball is rolling without slipping aross a horizontal table then its centre of mass must be moving in a straight line.

10. Suppose that a snooker ball is rolling without slipping on a horizontal table which is constrained to rotate about a vertical axis with constant speed Ω. The point O is the point of intersection of the axis with the plane of the table. Show that

$$\mathbf{v} + \boldsymbol{\omega} \wedge (-a\mathbf{k}) = \Omega\mathbf{k} \wedge \mathbf{r}$$

with the usual notation and with $\mathbf{r}$ being the position vector of the point of contact between the ball and the table, measured from O. Hence show that

$$\left(\frac{Ma^2}{A} + 1\right)\dot{\mathbf{r}} = \Omega\mathbf{k} \wedge \dot{\mathbf{r}}$$

where A is the moment of inertia about any axis through the centre of mass and the dot refers to rate of change in the inertial frame. Deduce that

$$\left(\frac{Ma^2}{A} + 1\right)\dot{\mathbf{r}} = \Omega\mathbf{k} \wedge (\mathbf{r} - \mathbf{c})$$

and hence

$$(\mathbf{r} - \mathbf{c})^2 = \text{constant}$$

for some constant vector $\mathbf{c}$. What is the path of the centre of mass in an inertial frame?

8 Lagrangian mechanics

8.1 Introduction

In previous chapters we have seen how the vector equations of motion are formulated both for particles and for rigid bodies. There are a limited number of examples in which a full analytic solution is possible. At some stage in the development of each solution it is necessary to think about a set of scalar coordinates which are appropriate to the geometry of the system under consideration. Sometimes this choice of coordinate system is clear cut. There is no need to use anything other than Cartesian corrdinates in the case of a projectile. On the other hand an angular coordinate is obviously a more suitable choice for a particle moving on a circular path. Consideration of the motion of a planet about a star is neatly tackled using plane polar coordinates given that such motion occurs in a plane. We should ask whether it is possible to cut out the initial stage of setting up the equations of motion using vectors and to move straight to the equations involving a chosen coordinate system. Lagrangian mechanics provides the means to do precisely that.

8.2 Lagrange's equations

In Chapter 6 we looked at the number of degrees of freedom of a mechanical system. It is worth repeating the definition here so that this section is complete in itself and easier to understand. It is a workable definition provided the constraints on the system are straighforward in character (see below).

Definition 8.1   The number of degrees of freedom of a mechanical system in an inertial frame is the number of independent scalar coordinates required to specify the positions of all particles in the system precisely at any given instant.

We have already noted that in general a rigid body has six degrees of freedom. A top spinning about its vertex which is free to move on a horizontal plane has five degrees of freedom. The top is constrained as its vertex is in contact with a plane. The position of the vertex could normally be described by three Cartesian coordinates in some inertial frame. Since it must remain on the plane the number of coordinates is reduced to two.

The other three degrees of freedom come from the rotational motion of the top as we have seen in Chapters 6 and 7.

If the plane has equation $ax + by + cz = d$ in the inertial frame then the coordinates of the vertex satisfy the equation of the plane. This type of constraint equation is called *holonomic*.

Definition 8.2

If a system is described by a set of scalar coordinates $q_1, q_2, q_3, \ldots, q_k$ then a constraint which can be written

$$\phi(q_1, q_2, \ldots, q_k, t) = 0$$

for some function ϕ is called *holonomic*. Any other type of constraint is called *non-holonomic*. If in addition ϕ is independent of t the constraint is *fixed* and holonomic.

We have met an example of a non-holonomic constraint before. In the general motion of a rolling sphere on a fixed horizontal plane the velocity of the point of contact is instantaneously zero, with constraint equation of the form

$$\mathbf{v} + \boldsymbol{\omega} \wedge \mathbf{r} = \mathbf{0}$$

(see Section 7.5). This is an equation involving velocities which in general cannot be integrated to give relations between the coordinates only. In the remainder of this chapter we shall consider only systems with *fixed holonomic* constraints.

In previous chapters we have shown how to calculate the kinetic energy for quite complicated mechanical systems. In general this function of coordinates and velocities is very much simpler to handle than the corresponding particle positions and velocities. It is also the case that for conservative systems the forces which do work are obtained from a scalar potential energy. Lagrange's analytical mechanics capitalizes on both these concepts. In order to set up the equations we shall need three definitions.

Definition 8.3

Coordinates $q_1, q_2, \ldots, q_n$ (not necessarily Cartesian) which specify a mechanical system completely are called *generalized coordinates*.

It follows that *if* all constraints are *holonomic* and *no subset of the coordinates* specifies the system completely then n is the *number of degrees of freedom* of the system.

Lagrange's equations for a mechanical system are set up with the aid of a mathematical device which is related to the geometry of the system but not to the actual motion. We consider a small displacement which is compatible with the fixed holomic constraints but which is not an actual displacement of the motion.

Defintion 8.4
A *virtual displacement* of a mechanical system (with purely fixed holonomic constraints) is a small displacement compatible with the constraints.

Example 1
Consider a simple pendulum consisting of a heavy bob with taut light inextensible string. Regarding the bob as a particle which is constrained to move in a fixed vertical plane through the point of suspension the position of the bob can be specified by the angle of inclination θ of the string to the downward vertical as shown in Fig. 8.1. The angle θ is then a generalized coordinate for the pendulum which has one degree of freedom.

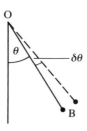

Fig. 8.1

A virtual displacement is represented by a small increase $\delta\theta$ in the angle.

Example 2
Consider the example above but with the string replaced by an elastic string. Then two coordinates are needed to describe the position of the system, the angle θ as above and the length of the string, say x. The system has two degrees of freedom and a virtual displacement consists of small changes $\delta\theta$ and δx in the two generalized coordinates.

Suppose that the system is formed for N particles each with position vector $\mathbf{r}_k$, $k = 1, 2, \ldots, N$. Then in order to enable us to use equations involving the generalized coordinates of the system $q_1, q_2, \ldots, q_n$ we must extend the idea of force.

Definition 8.5
The generalized forces Q_i are determined under a virtual displacement as follows:

$$\sum_{k=1}^{N} \mathbf{F}_k \cdot \delta\mathbf{r}_k = \sum_{i=1}^{n} Q_i \delta q_i \tag{8.1}$$

where $\mathbf{F}_k$ is the total force acting on the kth particle and $\delta\mathbf{r}_k$ is its displacement for the virtual displacement δq_i, $i = 1, 2, \ldots, n$. The left-hand sum is taken over the N particles.

Note that if n is the number of degrees of freedom then this equation determines the generalized forces in a straightforward fashion. Since $q_1, q_2, \ldots, q_n$ specify the position of the system we must have $\mathbf{r}_k = \mathbf{r}_k(q_1, q_2, \ldots, q_n)$ and by the chain rule the virtual displacement $\delta\mathbf{r}_k$ of the kth particle must satisfy

$$\delta\mathbf{r}_k = \sum_i \frac{\partial\mathbf{r}_k}{\partial q_i}\delta q_i \quad \text{for } k = 1, 2, \ldots, N. \tag{8.2}$$

Since δq_i are independent

$$Q_i = \sum_k \mathbf{F}_k \cdot \frac{\partial\mathbf{r}_k}{\partial q_i}. \tag{8.3}$$

We are now in a position to state and prove the theorem which gives Lagrange's equations in their general form for a system with fixed holonomic constraints.

Theorem 8.1   *Suppose that a mechanical system (with fixed holonomic constraints) is described by n independent generalized coordinates $q_1, q_2, \ldots, q_n$ with associated velocities $\dot{q}_1, \dot{q}_2, \ldots, \dot{q}_n$. Let $T = T(\mathbf{q}, \dot{\mathbf{q}})$ be the kinetic energy where $\mathbf{q}$ represents $q_1, q_2, \ldots, q_n$ and similarly $\dot{\mathbf{q}}$. Then the equations of motion of the system can be written as*

$$\frac{d}{dt}\left(\frac{\partial T}{\partial\dot{q}_i}\right) - \frac{\partial T}{\partial q_i} = Q_i \tag{8.4}$$

for $i = 1, 2, \ldots, n$, where Q_i are the generalized forces.

Proof   Consider the equation of motion for each particle in the system as derived form Newton's second law:

$$m_k\ddot{\mathbf{r}}_k = \mathbf{F}_k$$

where $\mathbf{F}_k$ is the total force acting on the kth particle. Taking the scalar product with $\delta\mathbf{r}_k$ and summing over the particles we have

$$\sum_k (m_k\ddot{\mathbf{r}}_k - \mathbf{F}_k)\cdot\delta\mathbf{r}_k = 0. \tag{8.5}$$

Consider the terms $\sum_k m_k\ddot{\mathbf{r}}_k \cdot \delta\mathbf{r}_k$. In order to avoid confusing notation we will drop the subscript k and write this as

$$\sum_{\text{part}} m\ddot{\mathbf{r}}.\delta\mathbf{r}$$

taking the sum over the particles as before.
   From (8.2) we have

$$\delta\mathbf{r} = \sum_i \frac{\partial\mathbf{r}}{\partial q_i}\delta q_i \tag{8.6}$$

and so

$$\sum_{\text{part}} m\ddot{\mathbf{r}}.\delta\mathbf{r} = \sum_{\text{part}}\sum_i m\ddot{\mathbf{r}}.\frac{\partial\mathbf{r}}{\partial q_i}\delta q_i. \tag{8.7}$$

Looking at the form of the equations (8.4) and remembering that $T = \sum_{\text{part}}\frac{1}{2}m\dot{\mathbf{r}}.\dot{\mathbf{r}}$, we investigate the coefficient of δq_i:

$$\sum_{\text{part}} m\ddot{\mathbf{r}}.\frac{\partial\mathbf{r}}{\partial q_i} = \sum_{\text{part}}\left[\frac{d}{dt}\left(m\dot{\mathbf{r}}.\frac{\partial\mathbf{r}}{\partial q_i}\right) - m\dot{\mathbf{r}}.\frac{d}{dt}\left(\frac{\partial\mathbf{r}}{\partial q_i}\right)\right]. \tag{8.8}$$

Now

$$\frac{d}{dt}\left(\frac{\partial\mathbf{r}}{\partial q_i}\right) = \sum_j \frac{\partial^2\mathbf{r}}{\partial q_i\partial q_j}\dot{q}_j = \frac{\partial}{\partial q_i}(\dot{\mathbf{r}}) \tag{8.9}$$

by the chain rule for partial differentiation.
   The equations regard T as a function of the $2n$ variables $q_1, q_2, \ldots, q_n$, $\dot{q}_1, \dot{q}_2, \ldots, \dot{q}_n$ which must be the case since

$$\dot{\mathbf{r}} = \sum_i \frac{\partial\mathbf{r}}{\partial q_i}\dot{q}_i \tag{8.10}$$

and

$$T = \frac{1}{2}\sum_{\text{part}} m\dot{\mathbf{r}}.\dot{\mathbf{r}} = \frac{1}{2}\sum_{\text{part}}\sum_i\sum_j m\frac{\partial\mathbf{r}}{\partial q_i}.\frac{\partial\mathbf{r}}{\partial q_j}\dot{q}_i\dot{q}_j. \tag{8.11}$$

Given that $\partial\mathbf{r}/\partial q_i$ is a function of $\{q_i\}$ only then (8.10) tells us that

$$\frac{\partial\dot{\mathbf{r}}}{\partial\dot{q}_i} = \frac{\partial\mathbf{r}}{\partial q_i}. \tag{8.12}$$

Putting the last four equations together in (8.8) we have

$$\sum_{\text{part}} m\ddot{\mathbf{r}}.\frac{\partial\mathbf{r}}{\partial q_i} = \sum_{\text{part}}\left[\frac{d}{dt}\left(m\dot{\mathbf{r}}.\frac{\partial\dot{\mathbf{r}}}{\partial\dot{q}_i}\right) - m\dot{\mathbf{r}}.\frac{\partial\dot{\mathbf{r}}}{\partial q_i}\right]$$

$$= \frac{d}{dt}\left(\frac{\partial T}{\partial\dot{q}_i}\right) - \frac{\partial T}{\partial q_i}. \tag{8.13}$$

Hence

$$\sum_{part} m\ddot{\mathbf{r}}.\delta\mathbf{r} = \sum_i \left[\frac{d}{dt}\left(\frac{\partial T}{\partial \dot{q}_i}\right) - \frac{\partial T}{\partial q_i} \right]\delta q_i$$

$$= \sum_{part} \mathbf{F}.\delta\mathbf{r} \qquad (8.14)$$

by Newton's second law for each particle (8.5). Using definition 8.5 for the generalized forces

$$\sum_{part} \mathbf{F}.\delta\mathbf{r} = \sum_i Q_i \delta q_i$$

and so eqn (8.5) can be written as

$$\sum_i \left[\frac{d}{dt}\left(\frac{\partial T}{\partial \dot{q}_i}\right) - \frac{\partial T}{\partial q_i} - Q_i \right]\delta q_i = 0. \qquad (8.15)$$

The small displacements δq_i, for $i = 1, 2, \ldots, n$, are independent so that we can set them all to zero with the exception of one. Consequently each coefficient of δq_i must be zero. We have

$$\frac{d}{dt}\left(\frac{\partial T}{\partial \dot{q}_i}\right) - \frac{\partial T}{\partial q_i} - Q_i = 0$$

and the theorem is proved.

The first advantage of this system of equations is that having selected a suitable set of generalized coordinates we need only find the kinetic energy, a scalar-valued function. The generalized forces can prove to be more of a problem but even here it turns out that simplifications follow. A very simple example helps us to see why.

Example 1 (cont.)   Consider the simple pendulum with one degree of freedom, as in example 1 above.

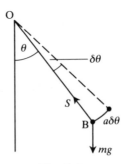

Fig. 8.2

There are two forces acting on the bob: the force due to gravity and the tension S in the string. Suppose that the length of the string is a. Then the kinetic energy of the bob is $T = \frac{1}{2}ma^2\dot{\theta}^2$. The work done by the two forces in a small virtual displacement $\delta\theta$ is

$$-mg\sin\theta\, a\delta\theta$$

involving only the force due to gravity, since the tension S in the string acts at right angles to the small displacement of the bob to first order in $\delta\theta$. The tension S is the constraint force, or more explicitly it is the force associated with the fact that the string is inextensible. The force of constraint does no work in the virtual displacement so that the generalized force Q_θ does not depend explicitly on it. We have

$$Q_\theta\,\delta\theta = -mga\sin\theta\,\delta\theta \Rightarrow Q_\theta = -mga\sin\theta.$$

Lagrange's equation reads

$$\frac{d}{dt}\left(\frac{\partial T}{\partial\dot{\theta}}\right) - \frac{\partial T}{\partial\theta} - Q_\theta = 0.$$

Using $T = \frac{1}{2}ma^2\dot{\theta}^2$

$$\frac{d}{dt}(ma^2\dot{\theta}) + mga\sin\theta = 0$$

giving the usual equation for a simple pendulum.

In so simple a system very little is gained by this method of approach. Even so we can note two points of interest.

1. The force associated with the holonomic constraint does no work in the virtual displacement and so never appears in the equation. This is useful since it is a force whose magnitude we would otherwise need to calculate.
2. This is a conservative system with a potential energy $V = -mga\cos\theta$ and we see that $Q_\theta = -\partial V/\partial\theta$.

Both points suggest that Lagrange's formulation should have a particular use in a conservative mechanical system. We make the following postulate:

d'Alembert's principle

In the virtual displacement of a mechanical system with purely holonomic constraints the forces of constraint do no work.

This postulate is generally easy to check in most systems. An example which makes the ideas clear is that of a particle sliding on a plane under the action of gravity. The normal reaction to the plane is the constraint force and since any virtual displacement of the particle occurs in the plane the work done by that force in such a virtual displacement is zero. *Note* that if the plane *is not smooth* there will be a frictional force which must do work and cannot be ignored, but that if the plane *is smooth* the system is conservative.

We will return to the general Lagrangian formulation and its application to a conservative system. Recall that a conservative mechanical system is one in which the forces which do work can each be derived from a scalar potential and that V, the potential energy of the system, is the sum of these potentials:

$$\mathbf{F}_k = -\operatorname{grad}\phi_k \quad \text{and} \quad V = \sum_k \phi_k.$$

Proposition 8.1   *For a conservative system with holonomic constraints the generalized forces Q_i are derived from the potential energy $V = V(q_1, q_2, \ldots, q_n)$ as follows*

$$Q_i = -\frac{\partial V}{\partial q_i}. \tag{8.16}$$

Proof   Assume that the force $\mathbf{F}_k$ acts at the point with position vector $\mathbf{r}_k$. Under a virtual displacement $\delta\mathbf{r}_k$ represents the displacement of this point of application. By the definition of generalized forces we have

$$\sum_k \mathbf{F}_k \cdot \delta\mathbf{r}_k = \sum_i Q_i \delta q_i \tag{8.17}$$

and also

$$\mathbf{F}_k \cdot \delta\mathbf{r}_k = -\operatorname{grad}\phi_k \cdot \delta\mathbf{r}_k = -\delta\phi_k.$$

The left-hand side of (8.17) is $-\sum \delta\phi_k = -\delta V$ by definition of V. But we know that $\phi_k = \phi_k(\mathbf{r}_k)$ and that $\mathbf{r}_k$ is a vector function of the generalized coordinates. Hence $V = \sum \phi_k = \sum \phi_k(q_1, q_2, \ldots, q_n) = V(q_1, q_2, \ldots, q_n)$ and by the chain rule

$$\delta V = \sum_i \frac{\partial V}{\partial q_i} \delta q_i$$

$$= -\sum_i Q_i \delta q_i$$

by (8.17). Again independence of δq_i gives

$$Q_i = -\frac{\partial V}{\partial q_i}.$$

The proposition is proved.

For the conservative system then Lagrange's equations become

$$\frac{d}{dt}\left(\frac{\partial T}{\partial \dot{q}_i}\right) - \frac{\partial T}{\partial q_i} + \frac{\partial V}{\partial q_i} = 0. \tag{8.18}$$

The new equations of motion become even simpler if we define a function $L(\mathbf{q}, \dot{\mathbf{q}})$. The single letter $\mathbf{q}$ represents $q_1, q_2, \ldots, q_n$, and similarly $\dot{\mathbf{q}}$ represents the generalized velocities $\dot{q}_1, \dot{q}_2, \ldots, \dot{q}_n$.

Definition 8.6

The Lagrangian L of a conservative mechanical system is defined to be

$$L = T - V$$

where T, V are the kinetic and potential energies respectively.

With this definition we get a particularly neat form of the equations.

Proposition 8.2

(*Lagrange's equations for a conservative system*) *If L is the Lagrangian of the system and q_i the generalized coordinates then the following equations hold:*

$$\frac{d}{dt}\left(\frac{\partial L}{\partial \dot{q}_i}\right) - \frac{\partial L}{\partial q_i} = 0 \quad \text{for } i = 1, 2, \ldots, n. \tag{8.19}$$

Proof

Equations (8.18) followed straight from the more general form of Lagrange's equations since $Q_i = -\partial V/\partial q_i$. We know that T is a function of both generalized velocities and coordinates whereas V is a function of the generalized coordinates alone. Hence $\partial L/\partial \dot{q}_i = \partial T/\partial \dot{q}_i$ and the theorem is proved.

We have set up the theory which we will need. We take the simplest case first and consider applications to two examples of motion of a single particle.

Example 2 (cont.)

This example involved a simple pendulum with the complication that the string holding the bob is elastic. There are two generalized coordinates, the angle θ of inclination of the string to the downward vertical and the length x of the string. Suppose that the elastic string satisfies Hooke's law with modulus of elasticity λ and natural length a. Then the tension S in the string satisfies

$$S = -\lambda \frac{(x - a)}{a}$$

and has associated potential $\frac{1}{2}\lambda(x - a)^2/a$ (half tension $\times$ extension). Adding in the potential energy due to gravity

$$V = -mga\cos\theta + \frac{1}{2}\lambda\frac{(x - a)^2}{a}.$$

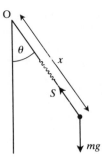

Fig. 8.3

The kinetic energy is given by

$$T = \tfrac{1}{2}m(\dot{x}^2 + x^2\dot{\theta}^2)$$

and so

$$L = \tfrac{1}{2}m(\dot{x}^2 + x^2\dot{\theta}^2) - \left(-mga\cos\theta + \frac{1}{2}\lambda\frac{(x-a)^2}{a}\right).$$

Using eqns (8.19) with $L = L(\theta, x, \dot{\theta}, \dot{x})$ we have

$$\frac{\mathrm{d}}{\mathrm{d}t}\left(\frac{\partial L}{\partial\dot{\theta}}\right) - \frac{\partial L}{\partial\theta} = 0 = \frac{\mathrm{d}}{\mathrm{d}t}(mx^2\dot{\theta}) - (-mga\sin\theta)$$

$$\frac{\mathrm{d}}{\mathrm{d}t}\left(\frac{\partial L}{\partial\dot{x}}\right) - \frac{\partial L}{\partial x} = 0 = \frac{\mathrm{d}}{\mathrm{d}t}(m\dot{x}) - \left(mx\dot{\theta}^2 - \lambda\frac{(x-a)}{a}\right).$$

We have the two equations

$$mx^2\ddot{\theta} + 2mx\dot{x}\dot{\theta} + mga\sin\theta = 0$$

$$m\ddot{x} - mx\dot{\theta}^2 + \lambda\frac{(x-a)}{a} = 0.$$

You can check that these equations agree with those obtained by using polar coordinates (x, θ) and Newton's second law for the bob.

Example 3   The other example we will consider is that of a particle sliding on the inside of a smooth bowl which has the form of a parabolic surface of revolution with axis of symmetry vertical. We use cylindrical polar coordinates (r, θ, z) where z is measured vertically upwards from the lowest point of the bowl. Suppose that the equation of the surface is given by $r^2 = az$.

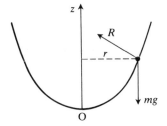

Fig. 8.4

The only forces are the force due to gravity and the normal reaction so that the system is conservative with potential energy mgz.

The kinetic energy in cylindrical polar coordinates is given by

$$T = \tfrac{1}{2}m(\dot{r}^2 + r^2\dot{\theta}^2 + \dot{z}^2).$$

However, we can only use Lagrange's equations if the generalized co-ordinates are independent. We must eliminate either r or z using the fixed holonomic constraint $r^2 = az$. Suppose that we decide to use r and θ as our independent coordinates. Then

$$L = T - V = \frac{1}{2}m\left(\dot{r}^2 + r^2\dot{\theta}^2 + 4\frac{r^2\dot{r}^2}{a^2}\right) - mg\frac{r^2}{a}.$$

Now we can use eqns (8.19) for both r and θ:

$$\frac{\mathrm{d}}{\mathrm{d}t}\left(\frac{\partial L}{\partial \dot{r}}\right) - \frac{\partial L}{\partial r} = \frac{\mathrm{d}}{\mathrm{d}t}\left(m\dot{r} + 4m\frac{r^2\dot{r}}{a^2}\right) - \left(mr\dot{\theta}^2 + 4m\frac{r\dot{r}^2}{a^2} - 2mg\frac{r}{a}\right) = 0$$

$$\frac{\mathrm{d}}{\mathrm{d}t}\left(\frac{\partial L}{\partial \dot{\theta}}\right) - \frac{\partial L}{\partial \theta} = \frac{\mathrm{d}}{\mathrm{d}t}(mr^2\dot{\theta}) = 0.$$

The last equation is particularly simple since L does not depend on θ. We can immediately integrate to give

$$mr^2\dot{\theta} = mh, \quad h \text{ constant.}$$

This is the familiar equation showing conservation of angular momentum about the vertical axis of symmetry. By substituting for $\dot{\theta}$ in terms of r we can find an equation in r alone describing the motion of the particle.

The example above shows how useful it can be if the Lagrangian L depends explicitly on $\dot{q}_j$ but not on q_j. In that case Lagrange's equation for the jth coordinate becomes

$$\frac{\mathrm{d}}{\mathrm{d}t}\left(\frac{\partial L}{\partial \dot{q}_j}\right) = 0. \tag{8.20}$$

This can be integrated immediately to give

$$\frac{\partial L}{\partial \dot{q}_j} = \text{constant.} \tag{8.21}$$

This type of coordinate proves so useful that it has merited a special name in the literature.

Definition 8.7   If the Lagrangian L does not depend explicitly on a generalized co-ordinate q_j but only on its generalized velocity $\dot{q}_j$ then q_j is said to be an *ignorable* coordinate and

$$\frac{\partial L}{\partial \dot{q}_j} = \text{constant.} \tag{8.22}$$

Both the examples 2 and 3 could be solved by using Newton's laws and/or conservation of energy. In the next two sections we shall see how Lagrange's equations provide an analytic framework for problems which are otherwise very difficult to solve. The theory developed in this chapter is of particular importance to the theory of small vibrations studied in the last section.

8.3   Application to rigid bodies

In Chapter 6 we have already learned that a rigid body in general motion has six degrees of freedom. What we have not yet covered is the choice of a convenient set of coordinates. This will have to be tackled if we are to use Lagrange's equations to describe the motion of a general rigid body. The vector equations are formulated in terms of the velocity of the centre of mass, **v**, and the angular velocity, **ω**. It is easy to choose three co-ordinates, which may well be Cartesian coordinates, in order to fix the position of the centre of mass. The other three degrees of freedom are related to the rotation of the body about its centre of mass, G. The coordinates used to describe this rotation are called Eulerian angles. They can be visualized as follows. The direction of a line through G and fixed in the body can be described in terms of spherical polar angles with respect to an inertial frame. This uses two angles θ, ϕ. Then the orientation of the body about this line requires a further angle ψ. The diagrams below (Figs 8.5–8.7) show how this is done in three steps introducing ϕ, θ, ψ in that order.

EULERIAN ANGLES

In the diagrams below $Oxyz$ is an inertial frame and G123 is a frame fixed relative to the body. Unit vectors along $Oxyz$ are written as $\mathbf{i}$, $\mathbf{j}$, $\mathbf{k}$ and those along G123 are $\mathbf{e}_1$, $\mathbf{e}_2$, $\mathbf{e}_3$. We find the directions of these last three vectors relative to $\mathbf{i}$, $\mathbf{j}$, $\mathbf{k}$ by three rotations through angles ϕ, θ, ψ, in that order as shown in the three diagrams, where $\mathbf{i}$, $\mathbf{j}$, $\mathbf{k}$ are drawn on the frame $Gxyz$ of parallel axes at G, since we are considering the rotation of the body about G.

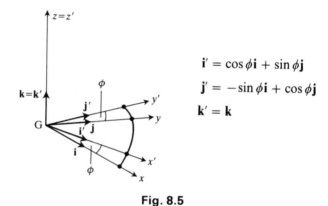

$$\mathbf{i}' = \cos\phi\,\mathbf{i} + \sin\phi\,\mathbf{j}$$
$$\mathbf{j}' = -\sin\phi\,\mathbf{i} + \cos\phi\,\mathbf{j}$$
$$\mathbf{k}' = \mathbf{k}$$

Fig. 8.5

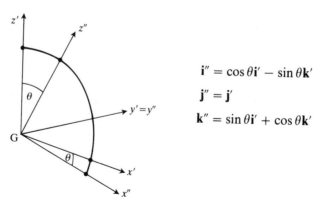

$$\mathbf{i}'' = \cos\theta\,\mathbf{i}' - \sin\theta\,\mathbf{k}'$$
$$\mathbf{j}'' = \mathbf{j}'$$
$$\mathbf{k}'' = \sin\theta\,\mathbf{i}' + \cos\theta\,\mathbf{k}'$$

Fig. 8.6

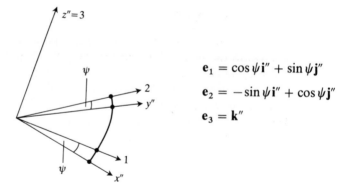

$$\mathbf{e}_1 = \cos \psi \mathbf{i}'' + \sin \psi \mathbf{j}''$$

$$\mathbf{e}_2 = -\sin \psi \mathbf{i}'' + \cos \psi \mathbf{j}''$$

$$\mathbf{e}_3 = \mathbf{k}''$$

Fig. 8.7

Putting these all together we have the following formulae for $\mathbf{e}_i$:

$$\mathbf{e}_1 = (\cos \theta \cos \phi \cos \psi - \sin \phi \sin \psi)\mathbf{i} + (\cos \theta \sin \phi \cos \psi + \cos \phi \sin \psi)\mathbf{j}$$
$$- \sin \theta \cos \psi \mathbf{k}$$

$$\mathbf{e}_2 = (-\cos \theta \cos \phi \sin \psi - \sin \phi \cos \psi)\mathbf{i}$$
$$+ (-\cos \theta \sin \phi \sin \psi + \cos \phi \cos \psi)\mathbf{j} + \sin \theta \sin \psi \mathbf{k}$$

$$\mathbf{e}_3 = \sin \theta \cos \phi \mathbf{i} + \sin \theta \sin \phi \mathbf{j} + \cos \theta \mathbf{k}. \tag{8.23}$$

It is now possible to see that $\mathbf{e}_3$ has spherical polar coordinates θ, ϕ with respect to $Gxyz$ and that ψ then represents a rotation of the body about the axis parallel to $\mathbf{e}_3$ through G as shown below.

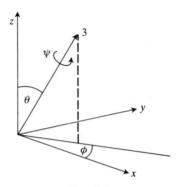

Fig. 8.8

Looking back at the three diagrams we can also find ω, the angular velocity, in terms of the Eulerian angles:

$$\omega = \dot{\phi}\mathbf{k} + \dot{\theta}\mathbf{j}' + \dot{\psi}\mathbf{e}_3. \tag{8.24}$$

We can collect all this together in terms of a very useful proposition.

Proposition 8.3 *Let* $\mathbf{e}_1, \mathbf{e}_2, \mathbf{e}_3$ *be unit vectors along a Cartesian frame of principal axes at G, fixed in the body, and with Eulerian angles* θ, ϕ, ψ *with respect to an inertial frame* $Oxyz$ *as above. Then the angular velocity of the body is given by*

$$\omega = (\dot{\theta}\sin\psi - \dot{\phi}\sin\theta\cos\psi)\mathbf{e}_1 + (\dot{\theta}\cos\psi + \dot{\phi}\sin\theta\sin\psi)\mathbf{e}_2$$
$$+ (\dot{\psi} + \dot{\phi}\cos\theta)\mathbf{e}_3 \tag{8.25}$$

and the angular momentum about G is

$$\mathbf{L} = A(\dot{\theta}\sin\psi - \dot{\phi}\sin\theta\cos\psi)\mathbf{e}_1 + B(\dot{\theta}\cos\psi + \dot{\phi}\sin\theta\sin\psi)\mathbf{e}_2$$
$$+ C(\dot{\psi} + \dot{\phi}\cos\theta)\mathbf{e}_3 \tag{8.26}$$

where A, B, C are the principal moments of inertia with respect to these axes at G.

Proof By (8.24) we have

$$\omega = \dot{\phi}\mathbf{k} + \dot{\theta}\mathbf{j}' + \dot{\psi}\mathbf{e}_3.$$

In order to use eqns (8.23) and Figs 8.5–8.7 above we note that

$$\mathbf{k} = (\mathbf{k}.\mathbf{e}_1)\mathbf{e}_1 + (\mathbf{k}.\mathbf{e}_2)\mathbf{e}_2 + (\mathbf{k}.\mathbf{e}_3)\mathbf{e}_3.$$

From (8.23) we have $\mathbf{k}.\mathbf{e}_1 = -\sin\theta\cos\psi$, $\mathbf{k}.\mathbf{e}_2 = \sin\theta\sin\psi$, and $\mathbf{k}.\mathbf{e}_3 = \cos\theta$. Also since $\mathbf{j}' = \mathbf{j}''$ Fig. 8.7 tells us that $\mathbf{j}' = \sin\psi\mathbf{e}_1 + \cos\psi\mathbf{e}_2$. These two results give (8.25) for ω. Since the axes G123 are principal axes in the body at G we know from Chapter 6 that

$$\mathbf{L} = (A\omega_1, B\omega_2, C\omega_3)$$

and hence (8.26) follows.

The key step in the use of Lagrange's equations lies in writing the kinetic energy in terms of the generalized coordinates and velocities. Having found the angular momentum we are in position to do just that.

Proposition 8.4 *If the centre of mass G has coordinates x, y, z with respect to an inertial frame and if principal axes at G are chosen with associated Eulerian angles* θ, ϕ, ψ *as above then the kinetic energy is given by*

$$T = \tfrac{1}{2}[A(\dot{\theta}\sin\psi - \dot{\phi}\sin\theta\cos\psi)^2 + B(\dot{\theta}\cos\psi + \dot{\phi}\sin\theta\sin\psi)^2$$
$$+ C(\dot{\psi} + \dot{\phi}\cos\theta)^2] + \tfrac{1}{2}M(\dot{x}^2 + \dot{y}^2 + \dot{z}^2). \qquad (8.27)$$

Proof   From (RB6) in Chapter 6 we have

$$T = \tfrac{1}{2}\boldsymbol{\omega}.\mathbf{L} + \tfrac{1}{2}M\mathbf{v}^2$$

and substituting the expressions for $\boldsymbol{\omega}$, $\mathbf{L}$ from proposition 8.3 the result follows.

**Corollary
8.4**   *In the special case where two of the principal moments of inertia are equal, $A = B$, then the kinetic energy takes on a much simplified form*

$$T = \tfrac{1}{2}[A\dot{\theta}^2 + A\dot{\phi}^2\sin^2\theta + C(\dot{\psi} + \dot{\phi}\cos\theta)^2] + \tfrac{1}{2}M(\dot{x}^2 + \dot{y}^2 + \dot{z}^2).$$
$$(8.28)$$

Following this corollary we can see that there will be an immediate application to an important example considered before in Chapter 7, that is the top and steady precession.

Example 4   *Steady precession of a top*

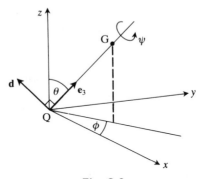

Fig. 8.9

Consider as in the diagram a top which is spinning with its vertex Q in contact with a rough horizontal floor so that the vertex does not move. Instead of using principal axes at the centre of mass it is more convenient to take axes at Q. Since $\mathbf{v}_Q = \mathbf{0}$ then

$$T = \tfrac{1}{2}\boldsymbol{\omega}.\mathfrak{I}_Q\boldsymbol{\omega}.$$

Suppose that the principal moments of inertia at Q are A, A, C and that we use Eulerian angles at Q as shown. Then the top has three degrees of freedom and by the same argument as for corollary 8.4 we have

$$T = \tfrac{1}{2}[A\dot{\theta}^2 + A\dot{\phi}^2 \sin^2\theta + C(\dot{\psi} + \dot{\phi}\cos\theta)^2]. \tag{8.29}$$

The angular velocity is

$$\boldsymbol{\omega} = \dot{\phi}\mathbf{k} + \dot{\theta}\mathbf{j}' + \dot{\psi}\mathbf{e}_3$$

where $\mathbf{k}$ is a unit vector vertically upwards, $\mathbf{e}_3$ is along the axis of symmetry, and $\mathbf{j}'$ is a horizontal vector perpendicular to the plane containing $\mathbf{e}_3$ and $\mathbf{k}$. The symmetry of the top leads to an *easier derivation of the kinetic energy*. As $\mathbf{j}'$ is perpendicular to the axis of symmetry it is in the direction of a principal axis at Q. We can choose principal axes $\mathbf{e}_3$, $\mathbf{j}'$, $\mathbf{d}$ at Q and then $\mathbf{d}$ is in the same plane as $\mathbf{e}_3$, $\mathbf{k}$. We have $\mathbf{k} = \cos\theta\mathbf{e}_3 + \sin\theta\mathbf{d}$ and so

$$\boldsymbol{\omega} = +\dot{\phi}\sin\theta\mathbf{d} + \dot{\theta}\mathbf{j}' + (\dot{\psi} + \dot{\phi}\cos\theta)\mathbf{e}_3 \tag{8.30}$$

giving

$$\mathbf{L}_Q = +A\dot{\phi}\sin\theta\mathbf{d} + A\dot{\theta}\mathbf{j}' + C(\dot{\psi} + \dot{\phi}\cos\theta)\mathbf{e}_3. \tag{8.31}$$

Knowing that $T = \tfrac{1}{2}\boldsymbol{\omega}.\mathfrak{I}_Q\boldsymbol{\omega}$ gives

$$T = \tfrac{1}{2}[A\dot{\theta}^2 + A\dot{\phi}^2 \sin^2\theta + C(\dot{\psi} + \dot{\phi}\cos\theta)^2] \tag{8.29}$$

as before.

The forces acting on the top are the reaction at Q, which includes the friction, and the force due to gravity at its centre of mass. Since the point Q is stationary by assumption we can see that the reaction does no work in any virtual displacement. The only other force is conservative and so the potential energy is

$$V = Mgh\cos\theta$$

and the Lagrangian becomes

$$L = \tfrac{1}{2}[A\dot{\theta}^2 + A\dot{\phi}^2 \sin^2\theta + C(\dot{\psi} + \dot{\phi}\cos\theta)^2] - Mgh\cos\theta \tag{8.32}$$

We can now write down Lagrange's equations in the three generalized coordinates θ, ϕ, ψ.

For ψ,

$$\frac{d}{dt}C(\dot{\psi} + \dot{\phi}\cos\theta) = 0. \tag{8.33}$$

For ϕ,

$$\frac{d}{dt}[A\dot{\phi}\sin^2\theta + C\cos\theta(\dot{\psi} + \dot{\phi}\cos\theta)] = 0. \tag{8.34}$$

For θ,

$$A\ddot{\theta} - [A\dot{\phi}^2 \sin\theta\cos\theta - C\dot{\phi}\sin\theta(\dot{\psi} + \dot{\phi}\cos\theta) + Mgh\sin\theta] = 0.$$
(8.35)

From eqns (8.33) and (8.34) we can see that ϕ, ψ are ignorable co-ordinates. Integrating these equations gives

$$\dot{\psi} + \dot{\phi}\cos\theta = n$$
(8.36)

where n is the constant spin of the top about its axis of symmetry and

$$A\dot{\phi}\sin^2\theta + C\cos\theta(\dot{\psi} + \dot{\phi}\cos\theta) = H$$

where H is a constant. This last equation is better rewritten as

$$A\dot{\phi}\sin^2\theta + Cn\cos\theta = H.$$
(8.37)

And from (8.35)

$$A\ddot{\theta} - [A\dot{\phi}^2\sin\theta\cos\theta - Cn\dot{\phi}\sin\theta + Mgh\sin\theta] = 0.$$
(8.38)

Equations (8.36), (8.37), (8.38) are the equations which we use to solve top problems with specific initial conditions.

ENERGY

In Chapter 7 we proved that the total energy $E = T + V$ is conserved. With a little manipulation the same deduction can be made from Lagrange's equations. Using (8.36) to remove $\dot{\psi}$ we have

$$E = \tfrac{1}{2}(A\dot{\theta}^2 + A\dot{\phi}^2\sin^2\theta + Cn^2) + Mgh\cos\theta.$$
(8.39)

In order to prove E is constant the simplest approach is to show that its rate of change with respect to time is zero:

$$\frac{dE}{dt} = A\dot{\theta}\ddot{\theta} + A\dot{\phi}\ddot{\phi}\sin^2\theta + A\dot{\phi}^2\dot{\theta}\sin\theta\cos\theta - Mgh\,\dot{\theta}\sin\theta.$$

But differentiating (8.37) gives

$$A\ddot{\phi}\sin^2\theta + 2A\dot{\phi}\dot{\theta}\sin\theta\cos\theta - Cn\dot{\theta}\sin\theta = 0$$

and substituting for $A\ddot{\phi}\sin^2\theta$ we have

$$\frac{dE}{dt} = A\ddot{\theta}\dot{\theta} + Cn\dot{\phi}\dot{\theta}\sin\theta - A\dot{\phi}^2\dot{\theta}\sin\theta\cos\theta - Mgh\dot{\theta}\sin\theta$$

$$= 0$$

by comparison with (8.38). As before we have shown that the energy is constant.

Eliminating $\dot{\phi}$ between (8.37) and (8.39) gives a first-order differential equation in θ which can be solved (numerically if necessary) in order to predict the behaviour of the top from known initial conditions.

PRECESSION

A top precesses as we saw in Chapter 7 if it inclines at a fixed angle α to the vertical, giving $\theta = \alpha$, $\dot{\theta} = 0$, and $\ddot{\theta} = 0$. Provided that $\alpha \neq 0$, from (8.37) it follows that $\dot{\phi}$ is constant, say Ω, and from (8.38) we have

$$0 - (A\Omega^2 \sin\alpha \cos\alpha - Cn\Omega \sin\alpha + Mgh \sin\alpha) = 0$$

and since $\alpha \neq 0$

$$A\Omega^2 \cos\alpha - Cn\Omega + Mgh = 0. \tag{8.40}$$

This gives immediately that precession implies steady precession, with constant angular speed of the axis of symmetry round the vertical, and also gives the condition (8.40) for steady precession to occur.

Example 5   (An ideal model of a penny spinning on a shiny table top.) Consider a uniform disc which is spinning freely and is moving on a smooth horizontal table, the rim of the disc being in contact with the table. The disc has mass m, radius a, and principal moments of inertia A, A, C at its centre of mass. If the disc is set spinning with spin Ω about a diagonal which is *almost* vertical show that

$$E = \frac{1}{2}\left((ma^2\cos^2\theta + A)\dot{\theta}^2 + \frac{A\Omega^2}{\sin^2\theta}\right) + mga\sin\theta = \tfrac{1}{2}A\Omega^2 + mga$$

where θ is the inclination of the normal to the disc to the upward vertical.

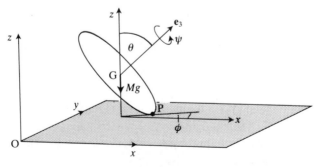

Fig. 8.10

The disc has five degrees of freedom since any rigid body has six degrees and the rim of the disc retains contact with the table. We will use Eulerian angles for the rotation about G as shown in Fig. 8.10. The position of G is fixed by two additional horizontal Cartesian coordinates since G is a height $a \sin \theta$ above the table. The disc is a top and so has a simplified form for the kinetic energy. With respect to axes $Oxyz$ fixed in the table the position of G is $(x, y, a \sin \theta)$. Using the general form for the kinetic energy

$$T = \tfrac{1}{2}m v_G^2 + \tfrac{1}{2}\boldsymbol{\omega}.\mathfrak{J}_G\boldsymbol{\omega}$$

$$T = \tfrac{1}{2}m(\dot{x}^2 + \dot{y}^2 + a^2\dot{\theta}^2 \cos^2\theta) + \tfrac{1}{2}[A\dot{\theta}^2 + A\dot{\phi}^2 \sin^2\theta + C(\dot{\psi} + \dot{\phi}\cos\theta)^2]$$

and $V = mga \sin \theta$, giving $L = T - V$, since the reaction at the point of contact is normal to the smooth table and does no work in a virtual displacement.

The initial conditions are: at $t = 0$, $\dot{x} = 0$, $\dot{y} = 0$, $\dot{\theta} = 0$, $\theta = \tfrac{1}{2}\pi$, $\dot{\phi} = \Omega$, $\dot{\psi} = 0$. There are five equations but four of them integrate immediately since x, y, ϕ, ψ are ignorable coordinates (L is a function of their generalized velocities only).

We must have $m\dot{x} = $ constant, $m\dot{y} = $ constant (integrating $\mathrm{d}/\mathrm{d}t(\partial T/\partial \dot{x}) = 0$ and similarly for y). Using the initial conditions $\dot{x} = 0 = \dot{y}$, we see that G does not move in a horizontal direction.

The equations involving ϕ and ψ are exactly the same as for the top and so we have

$$\dot{\psi} + \dot{\phi}\cos\theta = n \qquad A\dot{\phi}\sin^2\theta + Cn\cos\theta = H.$$

Using the initial conditions $n = 0$ since $\theta = \tfrac{1}{2}\pi$ and $\dot{\psi} = 0$. Then $H = A\Omega$ also. The equation involving θ gives

$$\frac{\mathrm{d}}{\mathrm{d}t}(ma^2\dot{\theta}\cos^2\theta + A\dot{\theta}) - [-ma^2\dot{\theta}^2 \cos\theta \sin\theta + A\dot{\phi}^2 \sin\theta \cos\theta$$

$$- C\dot{\phi}\sin\theta(\dot{\psi} + \dot{\phi}\cos\theta) - mga\cos\theta] = 0.$$

Using $n = 0$ and $A\dot{\phi}\sin^2\theta = A\Omega$ we have

$$(ma^2\cos^2\theta + A)\ddot{\theta} - ma^2\dot{\theta}^2 \sin\theta \cos\theta - A\Omega^2\left(\frac{\cos\theta}{\sin^3\theta}\right) + mga\cos\theta = 0. \tag{8.41}$$

But if we differentiate the expression for E given above in the example we get (8.41) multiplied by $\dot{\theta}$ and so $\mathrm{d}E/\mathrm{d}t = 0$ and E is constant. At $t = 0$, $\theta = \tfrac{1}{2}\pi$, $\dot{\theta} = 0$ and so $E = \tfrac{1}{2}A\Omega^2 + mga$.

At $t = 0$, putting $\theta = \tfrac{1}{2}\pi$, $\dot{\theta} = 0$ into (8.41) we have $\ddot{\theta} = 0$ also. Hence the disc only topples if the diagonal about which it is spinning is not quite vertical.

We can check whether the vertical position is stable by setting $\theta = \frac{1}{2}\pi + \varepsilon$ in (8.41). Approximating to first order in ε:

$$\sin(\tfrac{1}{2}\pi + \varepsilon) \approx 1 \qquad \cos(\tfrac{1}{2}\pi + \varepsilon) \approx -\varepsilon.$$

Equation (8.41) gives

$$A\ddot{\varepsilon} - 0 + A\Omega^2\varepsilon - mga\varepsilon \approx 0 \qquad (8.42)$$

correct to first order in ε and $\dot{\varepsilon}$. Comparison with the equation for simple harmonic motion

$$\ddot{\varepsilon} + \omega^2\varepsilon = 0$$

shows that the motion is stable for $A\Omega^2 - mga > 0$ and unstable otherwise.

This result gives an explanation for the way in which a penny behaves in contact with a table when the penny is spun about a vertical diagonal. There is a threshold for the spin below which the penny topples very quickly. For a large enough spin it takes some time to fall on to the table. The model above ignores friction at the point of contact but predicts accurately that the behaviour alters if the spin is big enough.

8.4 Small oscillations

Considering the last example it is clear that Lagrange's equations provide an ideal vehicle for studying the motion of a mechanical system about an equilibrium position (or about a relative equilibrium position as for the disc). A study of particular interest concerns the way in which a system behaves when slightly shifted from a position of stable equilibrium. In the case of the simple pendulum the bob oscillates with (angular) frequency $\sqrt{(g/a)}$ about its equilibrium position. For a compound pendulum (two bobs one below the other) there are two possible frequencies. These are the natural frequencies of the system. Finding them is of interest in itself for a general system and is of practical importance in a number of ways. If some external influence provides a forcing term which has a frequency close to one of the natural frequencies of the system a resonant response occurs and in physical structures this can do a great deal of damage. The old story that soldiers are ordered to break out of step when crossing a bridge has practical significance. A regular footfall could just possibly strike a natural frequency of the bridge.

In this section we will consider small oscillations about a position of stable equilibrium. The following assumptions apply throughout the section.

1. The generalized coordinates are chosen in such a way that the equilibrium position is given by $q_i = 0$ for all i.

2. The kinetic energy is quadratic in the velocities and does not depend explicitly on time.

3. The forces are conservative and given by a potential energy V which is a function of the generalized coordinates only.

4. The position of equilibrium is stable.

There are two stages to obtaining the possible frequencies for vibrations of the system. Firstly we find the approximate forms of the potential and kinetic energies. We can then deduce the form of the approximations to Lagrange's equations. We have to remember that as in simple harmonic motion with a single variable we are looking for a linear equation involving acceleration $\ddot{q}_i$ and generalized coordinates q_i.

Proposition 8.5   *Under the assumptions and correct to second order in the generalized coordinates and velocities, the kinetic and potential energies become*

$$T \approx \tfrac{1}{2}\dot{\mathbf{q}}^t T_0 \dot{\mathbf{q}} \quad \text{and} \quad V \approx \tfrac{1}{2}\mathbf{q}^t V_0 \mathbf{q} \tag{8.43}$$

where $\mathbf{q} = (q_1 q_2 \ldots q_n)^t$, *the superscript* t *denotes transpose (to avoid confusion with the kinetic energy), and* T_0, V_0 *are symmetric* $n \times n$ *matrices with constant entries.*

Proof   The kinetic energy is quadratic in the generalized velocities and can be written

$$T = \tfrac{1}{2} \sum_{i,j} a_{ij}(q_1, q_2, \ldots, q_n)\dot{q}_i \dot{q}_j$$

where we can choose the coefficients so that

$$a_{ij}(q_1, q_2, \ldots, q_n) = a_{ji}(q_1, q_2, \ldots, q_n).$$

Since each term is already second order in the velocities (which are zero in equilibrium) and since all the generalized coordinates are chosen so that $q_i = 0$ in equilibrium then each function $a_{ij}(q_1, q_2, \ldots, q_n)$ may be replaced by its value in equilibrium. Hence

$$T = \tfrac{1}{2} \sum_{i,j} a_{ij}(0, 0, \ldots, 0)\dot{q}_i \dot{q}_j.$$

Take $a_{ij} = a_{ij}(0, 0, \ldots, 0)$ and $T_0 = \{a_{ij}\}$. Then we have

$$T = \tfrac{1}{2}\dot{\mathbf{q}}^t T_0 \dot{\mathbf{q}}$$

where T_0 is a symmetric matrix as required, the superscript t denoting transpose.

To deal with V we must look at Lagrange's equations. These must be satisfied by $\mathbf{q} = 0$, $\dot{\mathbf{q}} = 0$, $\ddot{\mathbf{q}} = 0$ for all time t, since $\mathbf{q} = 0$ gives the equilibrium position. The equations are

$$\frac{d}{dt}\left(\frac{\partial T}{\partial \dot{q}_i}\right) - \frac{\partial T}{\partial q_i} + \frac{\partial V}{\partial q_i} = 0.$$

As T is quadratic in the velocities, when the substitutions $\dot{\mathbf{q}} = 0$, $\ddot{\mathbf{q}} = 0$ are made both terms involving T are zero. This gives

$$\frac{\partial V}{\partial q_i}(0,0,\ldots,0) = 0 \quad \text{for all } i = 1 \text{ to } n.$$

Now $V = V(q_1, q_2, \ldots, q_n)$ where q_i are small close to the equilibrium position. Using Taylor's expansion at the equilibrium position gives

$$V = V(q_1, q_2, \ldots, q_n) = V(0,0,\ldots,0) + \sum_i \frac{\partial V}{\partial q_i}(0,0,\ldots,0)q_i$$

$$+ \tfrac{1}{2}\sum_{i,j} \frac{\partial^2 V}{\partial q_i \partial q_j}(0,0,\ldots,0)q_i q_j + \text{higher terms.}$$

Since the potential energy is to be differentiated we may discard the constant term, $V(0,0,\ldots,0)$. We have just proved that all first-order partial derivatives are zero at the equilibrium position. Define

$$b_{ij} = \frac{\partial^2 V}{\partial q_i \partial q_j}(0,0,\ldots,0) \quad \text{and} \quad V_0 = \{b_{ij}\}.$$

Then V_0 is a symmetric matrix with constant coefficients and

$$V = \tfrac{1}{2}\mathbf{q}^t V_0 \mathbf{q}$$

as above.

Having determined the approximate form of the kinetic and potential energies we can derive Lagrange's equations in approximate form also.

Proposition 8.6 *Correct to first order in q_i, $\dot{q}_i$ Lagrange's equations become*

$$T_0 \ddot{\mathbf{q}} + V_0 \mathbf{q} = 0. \tag{8.44}$$

Proof The Lagrangian is given by

$$L \approx \tfrac{1}{2}\sum_{i,j} a_{ij}\dot{q}_i\dot{q}_j - \tfrac{1}{2}\sum_{ij} b_{ij}q_i q_j.$$

Using Lagrange's equation for q_i we have

$$\sum_j a_{ij}\ddot{q}_j + \sum_j b_{ij}q_j = 0 \quad \text{for } i = 1 \text{ to } n.$$

These n scalar equations are equivalent to the single matrix equation

$$T_0\ddot{\mathbf{q}} + V_0\mathbf{q} = \mathbf{0}$$

where $T_0 = \{a_{ij}\}$ and $V_0 = \{b_{ij}\}$.

These equations for $\mathbf{q}$ are a set of simultaneous linear differential equations with constant coefficients. The usual method of solution is to make a guess $\mathbf{q} = \mathbf{a}\exp(\lambda t)$ where $\mathbf{a}$, λ are constants. However, we know the equilibrium position is stable by assumption so that λ must be imaginary. We solve for $\mathbf{q}$ as usual but we make a more constructive guess.

Proposition 8.7   *The solutions of Lagrange's equations for small oscillations (8.44) are given by*

$$\mathbf{q} = \boldsymbol{\alpha}\cos(\omega t + \delta) \quad \text{where } (V_0 - \omega^2 T_0)\boldsymbol{\alpha} = \mathbf{0}. \tag{8.45}$$

Proof   We make the trial solution $\mathbf{q} = \boldsymbol{\alpha}\cos(\omega t + \delta)$ in eqn (8.44)

$$T_0\ddot{\mathbf{q}} + V_0\mathbf{q} = \mathbf{0}$$

giving

$$-\omega^2 T_0\boldsymbol{\alpha}\cos(\omega t + \delta) + V_0\boldsymbol{\alpha}\cos(\omega t + \delta) = \mathbf{0}.$$

Since $\cos(\omega t + \delta)$ is not zero for all t then

$$(V_0 - \omega^2 T_0)\boldsymbol{\alpha} = \mathbf{0}$$

and the proposition is proved.

Corollary 8.6   *The natural frequencies ω of the system satisfy the equation*

$$\det[V_0 - \omega^2 T_0] = 0. \tag{8.46}$$

The following definitions are in common use.

Definition 8.8   The solutions $\mathbf{q} = \boldsymbol{\alpha}\cos(\omega t + \delta)$ are called the *normal modes* of the system and the corresponding frequencies ω are called the *normal frequencies*.

The following proposition also holds. [You can find the proof in proposition 8.10 below.]

Proposition 8.8   *There are as many independent normal modes as there are degrees of freedom (under the assumptions previously made).*

Essentially when we find normal modes we are solving a set of eigenvector equations, the only difference being that T_0 replaces the more usual identity. The motion performed when the system is vibrating in a single normal mode is simple harmonic with a single frequency. The general solution for small oscillations is a combination of the normal modes

$$\mathbf{q} = \sum_i \alpha_i \cos(\omega_i t + \delta_i).$$

Example 6   (*The compound pendulum*) Consider two identical particles connected by light inextensible string of length a. One particle is attached by an identical string to a fixed point O. In equilibrium the particles hang with the two strings vertical. Find the normal frequencies and normal modes of the system if the particles are free to move in a vertical plane through 0.

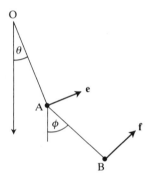

Fig. 8.11

In the diagram $\mathbf{e}$, $\mathbf{f}$ are unit vectors perpendicular to the strings OA, AB as shown. θ, ϕ are the inclinations of OA, AB to the vertical.

The first step is to find the kinetic and potential energies correct to second order. The velocity of particle A is $a\dot{\theta}\mathbf{e}$ and that of B is $a(\dot{\theta}\mathbf{e} + \dot{\phi}\mathbf{f})$:

$$T = \tfrac{1}{2}ma^2\dot{\theta}^2 + \tfrac{1}{2}ma^2(\dot{\theta}\mathbf{e} + \dot{\phi}\mathbf{f})^2$$
$$= \tfrac{1}{2}ma^2\dot{\theta}^2 + \tfrac{1}{2}ma^2[\dot{\theta}^2 + 2\dot{\theta}\dot{\phi}\cos(\theta - \phi) + \dot{\phi}^2].$$

We may use $\cos(\theta - \phi) \approx 1$ giving

$$T \approx ma^2\dot{\theta}^2 + ma^2\dot{\theta}\dot{\phi} + \tfrac{1}{2}ma^2\dot{\phi}^2.$$

Similarly

$$V = -mga\cos\theta - mg(a\cos\theta + a\cos\phi)$$
$$\approx \text{constant} + mga(\tfrac{1}{2}\theta^2 + \tfrac{1}{2}\theta^2 + \tfrac{1}{2}\phi^2).$$
$$T = \tfrac{1}{2}[\dot{\theta}\ \ \dot{\phi}]\begin{bmatrix} 2ma^2 & ma^2 \\ ma^2 & ma^2 \end{bmatrix}\begin{bmatrix} \dot{\theta} \\ \dot{\phi} \end{bmatrix}$$

and

$$V = \tfrac{1}{2}[\theta\ \ \phi]\begin{bmatrix} 2mga & 0 \\ 0 & mga \end{bmatrix}\begin{bmatrix} \theta \\ \phi \end{bmatrix}.$$

The 2×2 matrices are T_0 and V_0 respectively. We first solve the equation

$$\det[V_0 - \omega^2 T_0] = 0$$

$$\det\begin{bmatrix} 2mga - 2\omega^2 ma^2 & -\omega^2 ma^2 \\ -\omega^2 ma^2 & mga - \omega^2 ma^2 \end{bmatrix} = 0.$$

The simplest way of solving this equation is to divide every entry by mga and write $\omega^2 a/g = \lambda$, solving the resulting quadratic for λ. This gives

$$\det\begin{bmatrix} 2 - 2\lambda & -\lambda \\ -\lambda & 1 - \lambda \end{bmatrix} = 0 \Rightarrow \lambda = 2 \pm \sqrt{2}.$$

The two possibilities for the normal frequencies ω are

$$\omega^2 = \frac{g}{a}(2 \pm \sqrt{2}).$$

We can finish the example by solving for the normal modes.
   Solving for $\boldsymbol{\alpha} = [a_1\ \ a_2]^t$ we use

$$\begin{bmatrix} 2 - 2\lambda & -\lambda \\ -\lambda & 1 - \lambda \end{bmatrix}\begin{bmatrix} a_1 \\ a_2 \end{bmatrix} = \mathbf{0}.$$

Hence $\boldsymbol{\alpha} = A_1[1\ \ -\sqrt{2}]^t$ for $\lambda = 2 + \sqrt{2}$ and $\boldsymbol{\alpha} = A_2[1\ \ \sqrt{2}]^t$ for $\lambda = 2 - \sqrt{2}$. The normal modes are

$$\mathbf{q} = A_1[1\ \ -\sqrt{2}]^t \cos(\omega_1 t + \delta_1)$$

where $\omega_1^2 = (g/a)(2 + \sqrt{2})$ and

$$\mathbf{q} = A_2[1\ \ \sqrt{2}]^t \cos(\omega_2 t + \delta_2)$$

where $\omega_2^2 = (g/a)(2 - \sqrt{2})$.
   Notice that the normal mode in which the two particles move in opposition to each other has the higher frequency and the shorter period. This result seems very natural on physical grounds.

   The next example contains a useful method which gets around the difficulties caused by awkward choice of coordinates. The idea is to use

Cartesian coordinates if no other sensible set comes to mind. As the energy functions are only required to second order the coordinates can generally be reduced to the correct number of independent ones for the degrees of freedom involved. In order to consider an example which will be common to most people's experience we also need a result which will be proved in the next chapter and which is quoted here.

Proposition 8.9   *The kinetic energy of a uniform rod **AB** of mass m can be expressed as*

$$T = \frac{m}{6}(\mathbf{u}^2 + \mathbf{u}.\mathbf{v} + \mathbf{v}^2)$$

*where **u**, **v** are the velocities of the two ends of the rod. (NB Since the rod is of fixed length **u**, **v** have the same component along the rod.)*

Example 7   *(Swing)* Consider a uniform rod, mass m, length a, supported at each end by two light inextensible strings of length b. The strings are fastened to two points at the same vertical height and a distance a apart. Calculate the normal frequencies and the normal modes when the system performs small vibrations about the position of stable equilibrium.

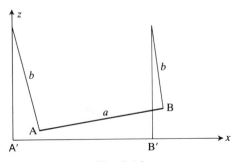

Fig. 8.12

We first note that in equilibrium the strings are vertical. Choose a set of Cartesian axes with the origin at A' (A in equilibrium position). A'x is then horizontal in the vertical plane through the rod in equilibrium, A'y is into the paper, and A'z is vertical. We take coordinates (x, y, z) for A and $(a + p, q, r)$ for B. Including a in the first coordinate for B ensures that the equilibrium position is given by all coordinates being zero. Then we can regard them all as small for our purpose. They cannot be independent as the strings and the rod have fixed lengths. We have the following constraints:

(C1) string at A $\Rightarrow$ $x^2 + y^2 + (b - z)^2 = b^2$

(C2) string at B $\Rightarrow$ $p^2 + q^2 + (b - r)^2 = b^2$

(C3) rod AB $\Rightarrow$ $(a + p - x)^2 + (q - y)^2 + (r - z)^2 = a^2$.

The respective expressions for the kinetic and potential energies are

(E1)
$$T = \frac{m}{6}(\dot{x}^2 + \dot{y}^2 + \dot{z}^2 + \dot{x}\dot{p} + \dot{y}\dot{q} + \dot{z}\dot{r} + \dot{p}^2 + \dot{q}^2 + \dot{r}^2)$$

using proposition 8.9, and

(E2)
$$V = \frac{mg}{2}(z + r).$$

The theory tells us that V should be second order in the coordinates. Looking at (C1) and (C2) gives

$$z = \frac{1}{2b}(x^2 + y^2 + z^2) \quad \text{and} \quad r = \frac{1}{2b}(p^2 + q^2 + r^2).$$

This means that r and z are of second order in the horizontal coordinates. We can write $V \approx (mg/4b)(x^2 + y^2 + p^2 + q^2)$ and also discard $\dot{z}^2$, $\dot{z}\dot{r}$, $\dot{r}^2$ from T.

We have reduced our six coordinates to four but one constraint remains. Expanding (C3) gives

$$2a(p - x) + (p - x)^2 + (q - y)^2 + (z - r)^2 = 0.$$

Correct to *first order* in all the coordinates we can see that $p \approx x$. The only terms left in T and V involving these two variables are squared terms and so we can replace p by x. We have

(E3)
$$T \approx \frac{m}{6}(3\dot{x}^2 + \dot{y}^2 + \dot{y}\dot{q} + \dot{q}^2) \quad \text{and} \quad V \approx \frac{mg}{4b}(2x^2 + y^2 + q^2).$$

The matrices we need are given by

$$T_0 = \frac{m}{6}\begin{bmatrix} 6 & 0 & 0 \\ 0 & 2 & 1 \\ 0 & 1 & 2 \end{bmatrix} \quad \text{and} \quad V_0 = \frac{mg}{2b}\begin{bmatrix} 2 & 0 & 0 \\ 0 & 1 & 0 \\ 0 & 0 & 1 \end{bmatrix}.$$

Using $\det[V_0 - \omega^2 T_0] = 0$ and setting $\lambda = b\omega^2/3g$ we have

$$\det\begin{bmatrix} 2 - 6\lambda & 0 & 0 \\ 0 & 1 - 2\lambda & -\lambda \\ 0 & -\lambda & 1 - 2\lambda \end{bmatrix} = 0.$$

This gives $\lambda = \frac{1}{3}$ or $3\lambda^2 - 4\lambda + 1 = 0 \Rightarrow \lambda = \frac{1}{3}, \frac{1}{3}, 1$.

The normal frequencies are given by

$$\omega = \sqrt{\frac{g}{b}} \text{ twice } \text{ and } \sqrt{\frac{3g}{b}}.$$

Solving the equation $[V_0 - \omega^2 T_0]\alpha = 0$ gives the following normal modes:

$$\mathbf{q} = A_1[1 \quad 0 \quad 0]^t \cos\left[\sqrt{\left(\frac{g}{b}\right)}t + \delta_1\right]$$

$$\mathbf{q} = A_2[0 \quad 1 \quad 1]^t \cos\left[\sqrt{\left(\frac{g}{b}\right)}t + \delta_2\right]$$

$$\mathbf{q} = A_3[0 \quad 1 \quad -1]^t \cos\left[\sqrt{\left(\frac{3g}{b}\right)}t + \delta_3\right].$$

The first two normal modes have exactly the same frequency as if the swing were a simple pendulum. One mode represents the swing going to and fro in the usual way; the other occurs as the swing moves from side to side in the vertical plane. In the last mode the y coordinates of the ends of the swing are equal and opposite. This gives a rotational motion. Again the larger frequency in this case seems natural due to twist caused in the ropes.

The final piece of theory in this chapter concerns a choice of coordinates which reduce a small oscillations problem to a standard form. We will study this approach briefly. There are two possible definitions of these coordinates. The two are almost equivalent!

Definition 8.9 (I) $Q_1, Q_2, \ldots, Q_n$ are called normal coordinates if

$$T = \tfrac{1}{2}(\dot{Q}_1^2 + \dot{Q}_2^2 + \cdots + \dot{Q}_n^2)$$
$$V = \tfrac{1}{2}(\omega_1^2 Q_1^2 + \omega_2^2 Q_2^2 + \cdots + \omega_n^2 Q_n^2) \qquad (8.47)$$

so that the normal coordinates simultaneously diagonalize both T and V.

Definition 8.9 (II) The normal coordinates are such that the normal modes are given by

$$Q_i = A_i \cos(\omega_i t + \delta_i) \qquad Q_j = \text{ for all } j \neq i$$

giving n modes as $i = 1$ to n.

If we use the first definition the kinetic energy matrix is the identity matrix, but the second definition only requires that both matrices are diagonal. The first definition implies the second. We will use the first definition.

The next proposition shows that every mechanical system oscillating about a stable equilibrium position has normal coordinates. *The proof demands a good knowledge of linear algebra and is not necessary in order to understand the remainder of the text.*

Proposition 8.10   *Normal coordinates always exist.*

Proof   We have $T = \frac{1}{2}\dot{\mathbf{q}}^t T_0 \dot{\mathbf{q}}$ and $V = \frac{1}{2}\mathbf{q}^t V_0 \mathbf{q}$. Since T is the kinetic energy whenever the system is in motion T is strictly positive. This tells us that the symmetric matrix T_0 is also positive with strictly positive eigenvalues. Using the theory of symmetric matrices we know $\exists P$ such that $P^t P = I$ and $P^t T_0 P = \mathrm{diag}(\lambda_1, \lambda_2, \ldots, \lambda_n) = D$ where $\lambda_i > 0 \; \forall i$. We can write $T_0 = PDP^t$ and $T = \frac{1}{2}\dot{\mathbf{q}}^t PDP^t \dot{\mathbf{q}}$. If we set $z = P^t \mathbf{q}$ then T becomes

$$T = \tfrac{1}{2}(\lambda_1 \dot{z}_1^2 + \lambda_2 \dot{z}_2^2 + \cdots + \lambda_n \dot{z}_n^2).$$

If we then let $\mathbf{y} = D^{1/2}\mathbf{z}$ so that $y_i = \sqrt{(\lambda_i)}z_i$ then we see that

$$T = \tfrac{1}{2}(\dot{y}_1^2 + \dot{y}_2^2 + \cdots + \dot{y}_n^2) = \tfrac{1}{2}\dot{\mathbf{y}}^t \dot{\mathbf{y}}$$

$$V = \tfrac{1}{2}\mathbf{y}^t D^{-1/2} P^t V_0 P D^{-1/2} \mathbf{y} \quad \text{since } \mathbf{q} = PD^{-1/2}\mathbf{y}.$$

If we set $B = D^{-1/2} P^t V_0 P D^{-1/2}$ then B is symmetric. This follows from the orthogonal property of P and the fact that $D^{-1/2}$ is diagonal and therefore symmetric. Again $\exists R$ such that $R^t R = I$ and $R^t B R = D_0$, a diagonal matrix. We make our final substitution of coordinates and set $\mathbf{Q} = R^t \mathbf{y}$ to give

$$V = \tfrac{1}{2}\mathbf{Q}^t D_0 \mathbf{Q}.$$

Since R is orthogonal T simply transforms to $T = \frac{1}{2}\dot{\mathbf{Q}}^t \dot{\mathbf{Q}}$ as required.

Assuming stable equilibrium this gives

$$T = \tfrac{1}{2}(\dot{Q}_1^2 + \dot{Q}_2^2 + \cdots + \dot{Q}_n^2)$$

$$V = \tfrac{1}{2}(\omega_1^2 Q_1^2 + \omega_2^2 Q_2^2 + \cdots + \omega_n^2 Q_n^2)$$

with associated Lagrange's equations

$$\ddot{Q}_i + \omega_i^2 Q_i = 0 \quad \text{for } i = 1 \text{ to } n.$$

METHOD USED TO CALCULATE THE NORMAL COORDINATES

The method is very much like that used to find the orthogonal matrix P which diagonalizes a symmetric matrix. We first solve the equations

(**) $$(V_0 - \omega^2 T_0)\boldsymbol{\alpha} = \mathbf{0}.$$

From the theorem above there must be n linearly independent solutions $\boldsymbol{\alpha}_1, \boldsymbol{\alpha}_2, \ldots, \boldsymbol{\alpha}_n$ with corresponding frequencies $\omega = \omega_i$.

Suppose that $\boldsymbol{\alpha}$ and $\boldsymbol{\beta}$ satisfy (**) with frequencies $\lambda^2 \neq \mu^2$. Then

$$(V_0 - \lambda^2 T_0)\boldsymbol{\alpha} = \mathbf{0} \quad \text{and} \quad \boldsymbol{\beta}^t(V_0 - \lambda^2 T_0)\boldsymbol{\alpha} = 0.$$

Taking the transpose

$$\boldsymbol{\alpha}^t(V_0 - \lambda^2 T_0)\boldsymbol{\beta} = 0$$

but $V_0\boldsymbol{\beta} = \mu^2 T_0 \boldsymbol{\beta}$. Hence

$$\boldsymbol{\alpha}^t(\mu^2 - \lambda^2)T_0\boldsymbol{\beta} = 0 \quad \text{and} \quad \mu^2 \neq \lambda^2$$

giving

$$\boldsymbol{\alpha}^t T_0 \boldsymbol{\beta} = \mathbf{0}.$$

Similarly

$$\boldsymbol{\alpha}^t V_0 \boldsymbol{\beta} = \mu^2 \boldsymbol{\alpha}^t T_0 \boldsymbol{\beta} = 0.$$

We then choose the length of each vector $\boldsymbol{\alpha}_i$ so that $\boldsymbol{\alpha}_i^t T_0 \boldsymbol{\alpha}_i = 1$. If there are repeated frequencies then there are as many linearly independent eigenvectors as there are repetitions of the frequency. These must be chosen so as to satisfy the equation below. In the case of different frequencies we have already proved that the equations hold with the correct choice of length for each $\boldsymbol{\alpha}_i$. Finally the eigenvectors chosen can be assumed to satisfy

$$\boldsymbol{\alpha}_i^t T_0 \boldsymbol{\alpha}_j = \delta_{ij} \quad \text{and} \quad \boldsymbol{\alpha}_i^t V_0 \boldsymbol{\alpha}_j = \omega_i^2 \delta_{ij}.$$

(Here $\delta_{ij} = 1$ if $i = j$ and $\delta_{ij} = 0$ if $i \neq j$.) Then we write $P = (\boldsymbol{\alpha}_1 \quad \boldsymbol{\alpha}_2 \quad \boldsymbol{\alpha}_n)$ putting the components of each vector $\boldsymbol{\alpha}_i$ in the columns of P. It follows from the equations above that

$$P^t T_0 P = I \quad \text{and} \quad P^t V_0 P = D_0 = \text{diag}(\omega_1^2, \omega_2^2, \ldots, \omega_n^2).$$

If we let $PQ = \mathbf{q}$ then

$$T = \tfrac{1}{2}\dot{\mathbf{q}}^t T_0 \dot{\mathbf{q}} = \tfrac{1}{2}\dot{Q}^t P^t T_0 P \dot{Q} = \tfrac{1}{2}\dot{Q}^t \dot{Q}$$

and similarly

$$V = \tfrac{1}{2}Q^t D_0 Q.$$

Since

$$P^t T_0 P = I \Leftrightarrow P^{-1} = P^t T_0$$

we can see that

$$Q = P^t T_0 \mathbf{q}.$$

Collecting all these results together we have proved the following.

Proposition   *If α_i are the eigenvectors satisfying (**) and if they are normalized so that*
8.11
$$\alpha_i^t T_0 \alpha_j = \delta_{ij}$$

then the matrix $P = (\alpha_1 \quad \alpha_2 \qquad \alpha_n)$ can be used to give the normal coordinates

$$\mathbf{Q} = P^t T_0 \mathbf{q}.$$

These results demand quite a good knowledge of algebra in order to prove them but are not so complicated in use in a particular example.

Example 6   In solving the oscillation problem for the compound pendulum we found
(cont.)   that for $\omega_1^2 = (g/a)(2 + \sqrt{2})$ then $\alpha_1 = A_1[1 \quad -\sqrt{2}]^t$ and for $\omega_2^2 = (g/a)(2 - \sqrt{2})$ then $\alpha_2 = A_2[1 \quad \sqrt{2}]^t$. Remember also that

$$T_0 = \begin{bmatrix} 2ma^2 & ma^2 \\ ma^2 & ma^2 \end{bmatrix}.$$

We need to choose A_1 and A_2 so that $\alpha_i^t T_0 \alpha_i = 1$. A little algebra gives

$$A_1 = [ma^2(4 - 2\sqrt{2})]^{-1/2} \qquad A_2 = [ma^2(4 + 2\sqrt{2})]^{-1/2}$$

and the matrix P becomes

$$P = \frac{1}{a\sqrt{m}} \begin{bmatrix} \dfrac{1}{\sqrt{(4 - 2\sqrt{2})}} & \dfrac{1}{\sqrt{(4 + 2\sqrt{2})}} \\[2ex] \dfrac{-\sqrt{2}}{\sqrt{(4 - 2\sqrt{2})}} & \dfrac{\sqrt{2}}{\sqrt{(4 + 2\sqrt{2})}} \end{bmatrix}.$$

The normal coordinates can then be found using

$$\begin{bmatrix} Q_1 \\ Q_2 \end{bmatrix} = P^t T_0 \begin{bmatrix} \theta \\ \phi \end{bmatrix}.$$

Finally

$$Q_1 = \frac{a\sqrt{m}}{\sqrt{(4 - 2\sqrt{2})}} [(2 - \sqrt{2})\theta + (1 - \sqrt{2})\phi]$$

$$Q_2 = \frac{a\sqrt{m}}{\sqrt{(4 + 2\sqrt{2})}} [(2 + \sqrt{2})\theta + (1 + \sqrt{2})\phi]$$

and you can check that these expressions give $T = \frac{1}{2}(\dot{Q}_1^2 + \dot{Q}_2^2)$.

In this chapter we have used Lagrange's equations to solve some old and new problems. Both the earlier vector methods used to express Newton's laws and the analytical methods of Lagrangian mechanics have their place. Lagrangian methods naturally lead to the development of mechanics on manifolds (in spaces which are not necessarily 'flat'). That step is beyond the scope of this book. We shall restrict ourselves to adapting both approaches to the study of impulsive motion in the final chapter.

Exercises:
Chapter 8

1. A satellite mass m is orbiting the Earth at some distance from it. Assume an inverse square law for the force towards the centre of the Earth and that the path is planar. Show that the Lagrangian can be written as

$$L = \tfrac{1}{2}m(\dot{r}^2 + r^2\dot{\theta}^2) + \frac{GMm}{r}$$

where r, θ are plane polar coordinates. Deduce that $r^2\dot{\theta} = h$, and write down Lagrange's equation in the coordinate r. Prove that the total energy is conserved.

2. Two small identical air pucks A, B are connected by a spring, spring constant k, natural length a, and are free to move along the straight line defined by the spring on the surface of a glass plate which is horizontal. Assuming that the frictional force between the pucks and the glass plate is negligible show that the Lagrangian is

$$L = \tfrac{1}{2}(m\dot{x}_1^2 + m\dot{x}_2^2) - \tfrac{1}{2}k(x_2 - x_1 - a)^2$$

where x_1, x_2 are measured form a fixed origin O on the line AB extended from A. Write down Lagrange's equations and show that the general solution takes the form

$$x_1 + x_2 = Ct + D$$

$$x_1 - x_2 = -a + A\cos\sqrt{\left(\frac{2k}{m}\right)}t + B\sin\sqrt{\left(\frac{2k}{m}\right)}t.$$

Describe the motion of the system.

3. A rod is constrained to move in a vertical plane. It is attached by a small smooth light ring at one of its ends to a smooth horizontal wire in the same plane. How many degrees of freedom has the rod?

   Find the Lagrangian of the system and reduce the problem to the solution of a second-order equation in θ, the angle between the rod and the downward vertical.

4. A small bead is sliding on a smooth vertical circular hoop of radius a, which is constrained to rotate with constant angular velocity ω about its vertical diameter. Using θ, the angle between the downward vertical and the radius to the bead, derive the Lagrangian $L(\theta, \dot{\theta})$ and show that if $g < a\omega^2$ the bead has four positions of equilibrium relative to the hoop whereas there are only two if $g > a\omega^2$. For what values of ω is the equilibrium position at the lowest point of the hoop stable?

   Why can we apply Lagrange's equations to this problem, thus ignoring the normal reaction between the hoop and the bead?

5. A spherical pendulum is formed from a light inextensible string of length a and a small heavy bob of mass m. Using spherical polars (θ, ϕ) with the polar

axis vertically downwards show that the Lagrangian is

$$L = \tfrac{1}{2}ma^2\dot\theta^2 + \tfrac{1}{2}ma^2\dot\phi^2 \sin^2\theta + mga\cos\theta.$$

Deduce that

$$ma^2\ddot\theta - ma^2\omega^2 \frac{\cos\theta}{\sin^3\theta} + mga\sin\theta = 0$$

where ω is a constant. Hence show that the total energy is conserved. Initially the bob is projected horizontally with velocity $\sqrt{(2)}a\Omega$ and with the string at an angle $\pi/4$ to the downward vertical. Assume that the string does not become slack. Show that the bob initially rises if $2\sqrt{(2)}a\Omega^2 > g$. How does the bob move if $2\sqrt{(2)}a\Omega^2 = g$?

6. A top is spinning about its vertex which is in contact with a very rough horizontal plane. The plane is rough enough to prevent the vertex from moving during the subsequent motion. Using the notation in the chapter, if the top is rotating with constant spin n about its axis of symmetry, which is vertical, and if it is then given a small push producing a small increment in $\dot\theta$, show that

$$A\ddot\theta + \frac{C^2n^2\sin\theta}{A(1+\cos\theta)^2} - Mgh\sin\theta = 0.$$

Deduce that the vertical position is stable provided the spin n satisfies

$$C^2n^2 > 4AMgh.$$

7. A particle A, mass m, is connected to a fixing point O by a light elastic string, modulus of elasticity λ and natural length a. The particle and string are free to move in a vertical plane containing O. Show that in small oscillations about the equilibrium position there are two possible normal frequencies, $\sqrt{(\lambda/ma)}$, $\sqrt{(g\lambda/a(\lambda + mg))}$, and describe the corresponding normal modes. What are the normal coordinates?

8. A pendulum consists of a light inextensible rod AB, length b, freely pivoted at A, attached at B to the rim of a uniform disc of radius a and mass m. The fixing between the disc and the rod has become loose so that the disc is free to rotate about B. The system moves in a vertical plane containing A. If θ is the angle between AB and the downward vertical, and ϕ is the angle between BG and the downward vertical, where G is the centre of mass of the disc, show that in small oscillations the kinetic and potential energies are given by

$$T \approx \tfrac{1}{2}mb^2\dot\theta^2 + mab\dot\theta\dot\phi + \tfrac{3}{4}ma^2\dot\phi^2$$
$$V = \tfrac{1}{2}mgb\theta^2 + \tfrac{1}{2}mga\phi^2.$$

Show that, if $a \ll b$, the normal frequencies are approximately

$$\sqrt{\left(\frac{g}{b}\right)} \quad \text{and} \quad \sqrt{\left(\frac{2g}{a}\left(1 + \frac{a}{b}\right)\right)}.$$

9. Prove that the stationary values and stationary points of the function

$$f(\mathbf{q}) = \frac{\mathbf{q}^t V_0 \mathbf{q}}{\mathbf{q}^t T_0 \mathbf{q}}$$

are the squared normal frequencies and are parallel to the normal modes, respectively, of the small oscillation problem with approximate Lagrangian

$$L = \tfrac{1}{2}(\dot{\mathbf{q}}^t T_0 \dot{\mathbf{q}} - \mathbf{q}^t V_0 \mathbf{q}).$$

10. A double swing is formed by two identical uniform rods AB, BC freely jointed at B. Each rod has mass m and length a. Three light ropes, length b, attach A, B, C to fixed points A′, B′, C′, which lie in a straight horizontal line with $A'B' = B'C' = a$. In equilibrium the ropes are vertical. How many degrees of freedom has the swing? Show that the system has the same normal frequencies as the single swing with the addition of a frequency $\sqrt{(3g/2b)}$. Find the corresponding normal modes and explain how the double swing moves in each mode.

9 Impulsive forces

9.1 Impulsive forces and Newton's laws

In the previous chapters we have dealt with continuously applied forces. We have not yet tackled the behaviour of mechanical systems which either suffer some sort of collision or which are subjected to a 'short sharp' blow. How do we describe mathematically the instantaneous effect of a blow from a hammer on a crowbar or the initial motion of a snooker ball after being struck by a cue? In both instances the time span of the blow (contact time) is so short as to be negligible, just a fraction of a second. In order to extend our model we must first state what we mean by an impulsive force.

Definition 9.1

An impulsive force is a force applied for an infinitesimally short time span.

Note that it is a vector quantity and that it is measured in newton seconds (Ns). In other words it is a very large force applied over a very short time span.

To establish the extension of Newtonian mechanics to cover impulsive forces and also collisions we set up new equations starting from the viewpoint that a large force is applied for a very short time. Then the *impulsive force* $\mathbf{J}$ is given by

$$\mathbf{J} = \lim_{\Delta t \to 0} \int_0^{\Delta t} \mathbf{F} \, dt \qquad (9.1)$$

where $\mathbf{F}$ is the force applied over the short time span Δt. Consider a particle moving in an inertial frame under the influence of such a force. Applying Newton's second law to the short interval over which the force is applied gives

$$m \frac{d\mathbf{v}}{dt} = \mathbf{F}$$

and integrating from 0 to Δt

$$m[\mathbf{v}(\Delta t) - \mathbf{v}(0)] = \int_0^{\Delta t} \mathbf{F} \, dt.$$

In the limit as $\Delta t \to 0$ this gives

$$m(\mathbf{v}_2 - \mathbf{v}_1) = \mathbf{J} \qquad (9.2)$$

where $\mathbf{v}_2$ is the velocity after the impulse and $\mathbf{v}_1$ is the velocity beforehand. We have proved the following.

Proposition 9.1 *If a particle is struck by an impulsive force $\mathbf{J}$ then the change in its momentum is given by*

$$m(\mathbf{v}_2 - \mathbf{v}_1) = \mathbf{J}.$$

In a collision between two particles the impulsive forces exerted by each on the other must be equal and opposite assuming Newton's third law. If the second particle is denoted by the primed variables and if the force exerted on the first particle by the second is $\mathbf{J}$ at the point of impact then we have

$$m(\mathbf{v}_2 - \mathbf{v}_1) = \mathbf{J}$$
$$m'(\mathbf{v}_2' - \mathbf{v}_1') = -\mathbf{J}.$$

Adding and rearranging gives

$$m\mathbf{v}_2 + m'\mathbf{v}_2' = m\mathbf{v}_1 + m'\mathbf{v}_1'. \qquad (9.3)$$

We can state this as

Proposition 9.2 *In a collision between particles, involving no other imulsive forces, the total linear momentum of the particles before the impact is equal to the total linear momentum after the impact.*

In such a collision the total linear momentum is conserved. The situation with regard to the kinetic energy is clear on physical grounds at least. In a straightforward collision between two particles with no other impulsive forces involved then at most the kinetic energy of the pair is preserved. Otherwise the total kinetic energy must decrease. For example if we treat the collision of two snooker balls as a collision of particles then a small amount of energy will be lost in sound as they collide.

Definition 9.2 A collision in which the total kinetic energy is preserved is said to be *elastic*. Any collision in which energy is dissipated is called inelastic.

This definition applies to particles and also to rigid bodies. (See exercises at the end of the chapter.)

The situation for particle collisions and impulses is now clear but since a particle has no shape or size this is an oversimplified model in most practical situations. The particle model gives us a start and enables us to formulate other models. One extension which we will briefly touch on here is the *variable mass equation*. Referring back to Chapter 1 and also using the ideas in this section, we can reformulate Newton's second law in such a way that it can be used in cases where the mass of the body concerned is not constant, for example in rocket motion.

APPLICATION TO ROCKET PROPULSION

Suppose that the particles in the exhaust gases of the rocket motor have velocity $\mathbf{u}(t)$ at time t relative to the rocket. Let the mass of the rocket at time t be $m(t)$ and its velocity $\mathbf{v}(t)$. To get a useful equation we must also include the action of a total external force $\mathbf{F}$ which can be a function of position, velocity, or time or any combination of the three. We will treat the rocket as though it were a heavy particle whose mass can change. Considering a small time interval $(t, t + \delta t)$ we can calculate the change of momentum in the rocket and exhaust gases combined. This change must be equal to the 'impulsive external force' applied, $\mathbf{F}\delta t$, reversing the argument in eqn (9.1). Over $(t, t + \delta t)$, eqn (9.2) gives (in an inertial frame)

$$m(t + \delta t)\mathbf{v}(t + \delta t) + [m(t) - m(t + \delta t)][\mathbf{v}(t + \delta t) - \mathbf{u}] - m(t)\mathbf{v}(t) \approx \mathbf{F}\delta t.$$

The middle term on the left is the approximate momentum of the exhaust gases in the inertial frame. Writing this rather more neatly and throwing away small terms gives

$$\delta m\mathbf{v} + m\delta\mathbf{v} - \delta m(\mathbf{v} - \mathbf{u}) \approx \mathbf{F}\delta t.$$

Dividing by δt and taking the limit as $\delta t \to 0$

$$m\frac{d\mathbf{v}}{dt} + \frac{dm}{dt}\mathbf{u} = \mathbf{F}.$$

Consider a situation where the rocket moves in a straight line, the exhaust speed of the gases relative to the rocket is u, and the rate of emission of exhaust gases is constant. Then $\dot{m} = -k$ for some constant k and the vector equation becomes a scalar equation

$$m\frac{dv}{dt} = F + ku.$$

The second term on the right indicates clearly how the exhaust gases provide the propulsion for the rocket.

The above equation assumes that the rocket can be represented as a particle of variable mass situated at its centre of mass. The exhaust gases can be emitted in a constant direction from the rocket given its long thin shape. However, imagine developing propulsion of a spherical rocket and controlling tendency to spin! Shape and inertia must be important.

RIGID BODY MOTION UNDER IMPULSIVE FORCES

The most interesting examples of impulsive motion, such as understanding how Uranus came to be spinning about an axis tipped at such an unusual angle relative to the plane containing its trajectory around the Sun, cannot be tackled by considering single particles. We need to develop the equations for the impulsive motion of a rigid body. Similarly if we are to consider the motion of a snooker ball we need to take account of the spin of the ball, as any player of the game knows. We will set up the equations from the equations for a particle as we did for ordinary motion under forces.

Proposition 9.3 *The change in the linear momentum of a rigid body acted on by impulsive forces is equal to the vector sum of the applied impulsive forces.*

Proof There are several ways of deriving this result. We will start with the individual particles forming the body. Then

$$m_i(\mathbf{v}_{i2} - \mathbf{v}_{i1}) = \sum \mathbf{J}_{ij} + \mathbf{J}_i \qquad (9.4)$$

using proposition 9.1 for the ith particle where $\mathbf{J}_{ij}$ is the impulsive force due to the jth particle and $\mathbf{J}_i$ is the applied impulsive force. Summing over the particles Newton's third law suggests that the internal forces cancel, $\mathbf{J}_{ij} = -\mathbf{J}_{ji}$, giving

$$\sum m_i(\mathbf{v}_{i2} - \mathbf{v}_{i1}) = \sum \mathbf{J}_i.$$

However, from the definition of the centre of mass G, if M is the mass of the body

$$M(\mathbf{v}_2 - \mathbf{v}_1) = \sum \mathbf{J}_i \qquad (9.5)$$

where $\mathbf{v}_2$ and $\mathbf{v}_1$ are the velocities of G before and after the impulse.

Proposition 9.4 *The change in the angular momentum of the body about* any *point Q is equal to the moment of the applied impulsive forces about Q.*

Proof For the ith particle taking the vector product of (9.4) with its position vector from Q

$$\mathbf{r}_i \wedge m_i(\mathbf{v}_{i2} - \mathbf{v}_{i1}) = \sum \mathbf{r}_i \wedge \mathbf{J}_{ij} + \mathbf{r}_i \wedge \mathbf{J}_i.$$

Summing over the particles and assuming that, as with forces applied over a longer time span, the combined moment of the internal impulsive forces is zero we have

$$\sum \mathbf{r}_i \wedge m_i(\mathbf{v}_{i2} - \mathbf{v}_{i1}) = \sum \mathbf{r}_i \wedge \mathbf{J}_i.$$

By definition of angular momentum at Q we have

$$\mathbf{L}_Q(2) - \mathbf{L}_Q(1) = \sum \mathbf{r}_i \wedge \mathbf{J}_i \qquad (9.6)$$

where the right-hand side consists of the moments of the external impulsive forces about Q and the left-hand side is the difference between the angular momentum at Q before and after the impulse.

This last equation shows an important difference when considered beside the usual equations involving rate of change of angular momentum. The equations derived in Chapter 6 can only be used at either the centre of mass or a fixed point Q, such that $v_Q = 0$ in the inertial frame. The impulsive angular momentum equation applies at any point Q, which need not be either the centre of mass or a fixed point. However, we have to be careful because it is still the case that only if Q = G or $v_Q = 0$ can we write

$$L_Q = \Im_Q \omega \tag{9.7}$$

where ω is the angular velocity and $\Im_Q$ is the inertia tensor at Q. For any other point we must use

$$L_Q = L_G + \underline{QG} \wedge M v_G \tag{9.8}$$

as in Chapter 6.

Propositions 9.3 and 9.4 have an important consequence which relates to the impact of two rigid bodies. Since the impulsive forces exerted by the bodies are equal and opposite by Newton's third law then in the absence of any other impulsive force we must have the following corollary.

Corollary   (*To propositions 9.3 and 9.4*) *The total linear momentum before and after the collision is the same* and *the total angular momentum about each point* P *is also conserved, irrespective of the choice of* P.

Throughout this section we have assumed that the impulsive forces are applied over such a short time span that they produce no change instantaneously in the position of the body, but do give a sudden jump in velocity (and angular momentum) and hence an infinite instantaneous acceleration.

We are now in a position to tackle some examples which occur in both everyday life and in astronomy.

Example 1   If a snooker player wishes to set the cue ball rolling without slipping how should he set about it?

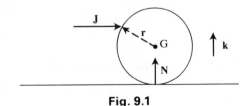

Fig. 9.1

We will assume that the ball is of uniform density and is perfectly spherical.

Which impulsive forces are involved? Over the instant of time for which the cue is in contact with the ball we assume it exerts an impulsive force **J** on the ball. In general motion there would be a normal reaction at the point of contact with the table, the weight acting vertically downwards, and the frictional force at the point of contact. Since the frictional force resists slippage at the point of contact our assumption is that there is no frictional *impulsive* force. Similarly the weight has no contribution to make over the instant of time of the impulse. However, there will be an impulsive normal reaction in general. Imagine the extreme case when the cue hits the top of the ball vertically downwards. Then there must be an impulsive reaction normal to the table at the point of contact. Suppose that the impulsive normal reaction is **N**.

We use the equation for change in linear momentum (9.5) to give

$$m\mathbf{v} = \mathbf{J} + \mathbf{N} \tag{9.9}$$

where **v** is the velocity of the centre of mass G after the blow. Taking the vertical component

$$m\mathbf{v}.\mathbf{k} = \mathbf{J}.\mathbf{k} + N$$

If $\mathbf{J}.\mathbf{k} > 0$ (difficult to achieve with a snooker cue!) then since $N \geq 0$ we must have the ball leaving the table. If $\mathbf{J}.\mathbf{k} < 0$ the player wastes some of his effort. Hence the most efficient approach is to get the cue as near to horizontal as possible. Assume that $\mathbf{J}.\mathbf{k} = 0$ so that **J** is horizontal. Then (9.9) gives

$$m\mathbf{v} = \mathbf{J}. \tag{9.10}$$

We can then use the equation for change of angular momentum at G:

$$\mathbf{L}(G) = \mathbf{r} \wedge \mathbf{J}$$

where **r** is the position vector from G of the point at which the cue strikes. At the centre of mass we have $\mathbf{L}(G) = \frac{2}{5}ma^2\boldsymbol{\omega}$ giving

$$\tfrac{2}{5}ma^2\boldsymbol{\omega} = \mathbf{r} \wedge \mathbf{J}. \tag{9.11}$$

Between (9.10) and (9.11) we have

$$\boldsymbol{\omega} = \frac{5}{2a^2}\mathbf{r} \wedge \mathbf{v}.$$

If the initial motion of the ball is to roll without slipping then the initial velocity of the point of contact must be zero, giving

$$\mathbf{0} = \mathbf{v} + \boldsymbol{\omega} \wedge (-a\mathbf{k}).$$

If this is satisfied then

$$0 = \mathbf{v} - \frac{5}{2a}(\mathbf{r} \wedge \mathbf{v}) \wedge \mathbf{k}$$

or

$$0 = \mathbf{v} - \frac{5}{2a}(\mathbf{r} \cdot \mathbf{k})\mathbf{v} \qquad (9.12)$$

since $\mathbf{v} \cdot \mathbf{k} = 0$. Equation (9.12) shows that to achieve no slippage at the point of contact we must have

$$(\mathbf{r} \cdot \mathbf{k}) = \tfrac{2}{5}a.$$

The cue must strike the ball at any point $\tfrac{7}{5}a$ above the table where a is the radius of the ball. If the ball is hit above or below this level it has either 'top' spin or 'back' spin respectively. (See exercise at the end of the chapter.)

Example 2   Unlike the Earth the axis of rotation of the planet Uranus about its own centre of mass is inclined at a very pronounced angle to the normal to the plane in which Uranus moves about the Sun. It is thought that this could be due to collision with a heavy fast-moving meteorite. We can look at this suggestion using the theory developed above. Let us start with a planet spinning about a polar axis perpendicular to its plane of motion about the Sun. A simplifying assumption which we will make is that the meteorite stays put on impact so that its constituent material adheres to the surface of the planet. This eases the calculations without rendering them useless! We will model the meteorite as a particle.

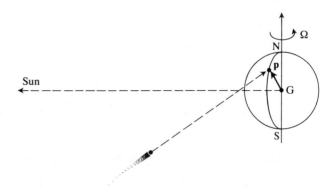

Fig. 9.2

In the above diagram $\boldsymbol{\Omega}$ is the spin of the planet about its polar axis before collision and $\mathbf{V}$ is its velocity (taken to be perpendicular to $\boldsymbol{\Omega}$ and roughly into the paper). Suppose that the mass of the planet is M, that of the particle is m, and that the particle makes contact with the planet at a point on its surface position vector $\mathbf{p}$ from the centre of mass. We will assume that the planet is spherical and that its moment of inertia is C.

The corollary to propositions 9.3 and 9.4 above tells us that linear momentum and angular momentum about the centre of mass are conserved. We have the equations

$$M\mathbf{V} + m\mathbf{v} = M\mathbf{u} + m(\mathbf{u} + \boldsymbol{\omega} \wedge \mathbf{p}) \tag{9.13}$$

$$C\boldsymbol{\Omega} + \mathbf{p} \wedge m\mathbf{v} = C\boldsymbol{\omega} + \mathbf{p} \wedge m(\mathbf{u} + \boldsymbol{\omega} \wedge \mathbf{p}) \tag{9.14}$$

since immediately after the collision the meteorite has velocity $\mathbf{u} + \boldsymbol{\omega} \wedge \mathbf{p}$, the same velocity as the point of impact on the planet to which it adheres. We are interested in calculating $\boldsymbol{\omega}$, the final angular velocity, and so must eliminate $\mathbf{u}$, the velocity of G, between the two equations. However, we could assume as a sensible approximation that both the terms containing $m(\mathbf{u} + \boldsymbol{\omega} \wedge \mathbf{p})$ in (9.13) and (9.14) are small. In other words we assume that $m \ll M$ but that $\mathbf{v}$ is large in comparison with $\mathbf{u}$, $\mathbf{V}$ and $\boldsymbol{\omega} \wedge \mathbf{p}$. Then approximately

$$M\mathbf{V} + m\mathbf{v} \approx M\mathbf{u} \tag{9.15}$$

$$C\boldsymbol{\Omega} + \mathbf{p} \wedge m\mathbf{v} \approx C\boldsymbol{\omega}. \tag{9.16}$$

If $\mathbf{v}$ is in the same plane as the path of the planet, eqn (9.15) suggests this plane remains the same (to our level of approximation). Equation (9.16) can be rewritten as

$$\boldsymbol{\omega} \approx \boldsymbol{\Omega} + C^{-1}\mathbf{p} \wedge m\mathbf{v}. \tag{9.17}$$

If the meteorite hits the planet close to one of the poles, say N on the diagram, then $\mathbf{p}$ is very close to being parallel to $\boldsymbol{\Omega}$ and the second term in (9.17) can produce a substantial component perpendicular to $\boldsymbol{\Omega}$. This suggests that a reasonably heavy meteorite travelling at high speed could substantially alter the orientation of the polar axis. The idea that the axis of rotation of the planet could have been altered due to a collision is certainly a possible one.

9.2 Extension of Largrange's equations to impulsive motion

One of the important applications of Lagrange's equations is to impulsive motion. We can move from the vector formulation of Newton's laws to Lagrangian methods as in the case of ordinary motion. Firstly consider Lagrange's equations for a normal mechanical system with generalized coordinates $q_1, q_2, \ldots, q_n$:

$$\frac{\mathrm{d}}{\mathrm{d}t}\left(\frac{\partial T}{\partial \dot{q}_i}\right) - \frac{\partial T}{\partial q_i} = Q_i \tag{9.18}$$

where the generalized forces Q_i are given by

$$\sum \mathbf{F}.\delta\mathbf{r} = \sum_i Q_i\delta q_i. \qquad (9.19)$$

The sum on the left is the sum over all forces contributing to the motion of the mechanical system. In order to apply these to an impulsive motion we should note that in the previous section we assumed: (i) impulsive forces are very large forces applied over a very short period of time; (ii) the result is an instantaneous jump in the velocities but no instantaneous change in position.

Proposition 9.5 *Lagrange's equations for impulses can be expressed in the form*

$$\left[\frac{\partial T}{\partial \dot{q}_i}\right]_1^2 = \hat{Q}_i \qquad (9.20)$$

where

$$\sum \mathbf{J}.\delta\mathbf{r} = \sum_i \hat{Q}_i\delta q_i \qquad (9.21)$$

each $\mathbf{J}$ *being an impulsive force and* $\hat{Q}_i$ *being a generalized impulsive force.*

Proof Integrate eqn (9.18) over the very small time δt in which the impulsive force is applied:

$$\left[\frac{\partial T}{\partial \dot{q}_i}\right]_0^{\delta t} - \int_0^{\delta t}\frac{\partial T}{\partial q_i}\,dt = \int_0^{\delta t} Q_i\,dt. \qquad (9.22a)$$

We take the limit as $\delta t \to 0$ and define

$$\hat{Q}_i = \lim \int_0^{\delta t} Q_i\,dt. \qquad (9.22b)$$

On the left-hand side of (9.22a) the first term gives the change in the partial derivatives of the kinetic energy due to the sudden change in the generalized velocities. The second term becomes tiny

$$\int_0^{\delta t}\frac{\partial T}{\partial q_i}\,dt \to 0.$$

To see this, note that the integrand is a function of generalized velocities and coordinates only. It contains no accelerations. The generalized coordinates remain the same and whilst the generalized velocities jump in value they are nevertheless finite. This means that the function $\partial T/\partial q_i$ is bounded and as $\delta t \to 0$ the integral must tend to zero. It remains to establish the links with the impulsive forces. Remember that the virtual displacements are independent of the true motion and of time. Take eqn (9.19) and integrate:

$$\sum \int_0^{\delta t} \mathbf{F}\,dt.\delta\mathbf{r} = \sum_i \int_0^{\delta t} Q_i\,dt\,\delta q_i.$$

By definition

$$\int_0^{\delta t} \mathbf{F}\, dt \to \mathbf{J} \quad \text{and} \quad \int_0^{\delta t} Q_i\, dt \to \hat{Q}_i$$

and so (9.21) is proved.

  In solving these equations it is generally the case that we are concerned only with velocities and so do not need to look at the generalized coordinates. To make the calculation of $\hat{Q}_i$ easier we will adapt (9.21) for use with velocities.

Corollary 9.5

$$\sum \mathbf{J}.\mathbf{v} = \sum_i \hat{Q}_i u_i \tag{9.23}$$

where $\mathbf{v}$ is a velocity of the point of contact of the impulse $\mathbf{J}$ which results from arbitrary instantaneous velocities $\dot{q}_i = u_i$ (compatible with the constraints).

Proof

If $\mathbf{r}(q_1, q_2, \dots, q_n)$ is the position vector of the point of application of the impulsive force $\mathbf{J}$ then the chain rule gives

$$\delta \mathbf{r} = \sum_i \frac{\partial \mathbf{r}}{\partial q_i} \delta q_i.$$

If we generate a virtual displacement by letting $\delta q_i = u_i \delta t$ where u_i is an arbitrary constant then

$$\delta \mathbf{r} = \mathbf{v} \delta t = \sum_i \frac{\partial \mathbf{r}}{\partial q_i} u_i \delta t.$$

If $q_1, q_2, \dots, q_n$ are independent coordinates then $u_1, u_2, \dots, u_n$ are independent also. Replacing $\delta \mathbf{r}$ and δq_i in (9.21) gives

$$\sum \mathbf{J}.\mathbf{v} \delta t = \sum_i \hat{Q}_i u_i \delta t.$$

Since $u_1, u_2, \dots, u_n$ are independent we must be able to determine $\hat{Q}_i$ uniquely from these equations as before.

  In order to simplify the notation we can write $\dot{q}_i = u_i$ in eqns (9.20) giving

$$\left[\frac{\partial T}{\partial u_i} \right]_1^2 = \hat{Q}_i. \tag{9.24}$$

The equations developed above are extremely simple to use in examples where the choice of independent scalar velocities is clear. Problems involving rods are particularly easy to solve provided that the formula below for calculating the kinetic energy of a rod is used.

Proposition 9.6 *A thin uniform rod has kinetic energy*

$$\tfrac{1}{6}m(u_A^2 + u_B^2 + \mathbf{u_A}\cdot\mathbf{u_B}) \tag{9.25}$$

where m is the mass and $\mathbf{u_A}$ *and* $\mathbf{u_B}$ *are the velocities of the ends of the rod.*

Proof If $\mathbf{v}$ is the velocity of the centre of mass G and ω is the angular velocity (assumed perpendicular to the rod) then the kinetic energy of the rod is given by

$$T = \tfrac{1}{2}mv^2 + \tfrac{1}{2}\omega.\mathbf{L_G}$$
$$= \tfrac{1}{2}mv^2 + \tfrac{1}{6}ma^2\omega^2$$

where $2a$ is the length of the rod and ω is perpendicular to the rod. Using the usual formula connecting velocities of two points in a rigid body we have

$$\mathbf{u_A} = \mathbf{v} + \omega \wedge (-a\mathbf{e})$$
$$\mathbf{u_B} = \mathbf{v} + \omega \wedge a\mathbf{e}$$

if $\mathbf{e}$ is a unit vector in the direction AB. These equations give

$$\tfrac{1}{2}(\mathbf{u_A} + \mathbf{u_B}) = \mathbf{v} \quad \text{and} \quad \tfrac{1}{2}(\mathbf{u_B} - \mathbf{u_A}) = \omega \wedge a\mathbf{e}.$$

ω and $\mathbf{e}$ are orthogonal vectors and so the length of $\omega \wedge a\mathbf{e}$ is just $a|\omega|$. Substitution in the expression for T gives

$$T = \tfrac{1}{8}m(\mathbf{u_A} + \mathbf{u_B})^2 + \tfrac{1}{24}m(\mathbf{u_A} - \mathbf{u_B})^2$$

and the result is proved.

Example 3 Four uniform identical rods are freely jointed at their ends A, B, C, D. They are lying at rest on a table in the shape of a square when an impulsive force J is applied at A towards C. Show that the distance between A and C initially begins to decrease.

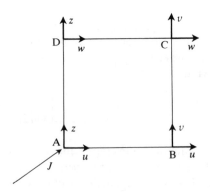

Fig. 9.3

This is essentially a two-dimensional problem only. We choose velocities u, v, w, z as shown in the diagram. We only need to satisfy the constraints that the lengths of the rods are preserved at the instant of the blow. This means for example that the component of velocity of A parallel to the rod AB is the same as the component of velocity of B in that direction. If that were not the case then the rod would be either lengthening or shortening. Then the kinetic energy of the rod AB is

$$\tfrac{1}{6}m(3u^2 + z^2 + v^2 + zv)$$

using the formula for the kinetic energy of a rod AB as in (9.25).
    Hence the total kinetic energy for all four rods is

$$T = \tfrac{1}{6}m(5u^2 + 5v^2 + 5w^2 + 5z^2 + 2uw + 2vz).$$

If each rod moves with velocity as shown then eqn (9.23) gives

$$\frac{J}{\sqrt{2}}(u + z) = J_u u + J_v v + J_w w + J_z z.$$

Then Lagrange's equations for impulse give

$$\left[\frac{\partial T}{\partial u}\right]_1^2 = J_u \Rightarrow \tfrac{1}{6}m(10u + 2w) = \frac{1}{\sqrt{2}} J.$$

For v

$$\tfrac{1}{6}m(10v + 2z) = 0.$$

For w

$$\tfrac{1}{6}m(10w + 2u) = 0.$$

For z

$$\tfrac{1}{6}m(10z + 2v) = \frac{1}{\sqrt{2}} J.$$

These equations are easily solved to give

$$u = z = \frac{5J}{8\sqrt{(2)}m} \qquad v = w = -\frac{J}{8\sqrt{(2)}m}.$$

From these we see that the component of velocity of A in the direction AC is

$$\frac{1}{\sqrt{2}}(u + z) = \frac{5J}{8m}$$

whereas that of C is

$$\frac{1}{\sqrt{2}}(v + w) = -\frac{J}{8m}$$

making it clear that AC initially starts to shorten the initial rate of change of length being $-6J/8m$.

If in addition we wished to determine the kinetic energy after the impulse then we could substitute for the velocities in the expression for the kinetic energy. However, that leaves a rather messy calculation. There is a neater formula for systems starting from rest. (This can be extended to the general case.)

Proposition 9.7

If the kinetic energy of the system is quadratic (homogeneous of degree 2) in the velocities and if the system starts from rest then the kinetic energy after an impulse is given by

$$T = \tfrac{1}{2} \sum \hat{Q}_i u_i \tag{9.26}$$

where u_i is the generalized velocity immediately after the impulse.

Proof

Since the system starts from rest the velocities are all zero before the impulse.

The kinetic energy is homogeneous and so

$$2T = \sum \frac{\partial T}{\partial u_i} u_i.$$

Equation (9.24) becomes

$$\frac{\partial T}{\partial u_i} = \hat{Q}_i \quad \text{and so} \quad 2T = \sum \hat{Q}_i u_i.$$

Example 3 (cont.)

To calculate the kinetic energy for the four rods immediately after the impulse we use (9.26):

$$T = \frac{1}{2} \frac{J}{\sqrt{2}} (u + z) = \frac{1}{2} \frac{J}{\sqrt{2}} \left(\frac{5J}{8\sqrt{(2)}m} 2 \right) = \frac{5J^2}{16m}.$$

Example 4

A top, with fixed vertex P, is performing steady precession inclined at an angle α to the upward vertical with spin n about its axis of symmetry which has angular velocity Ω about the vertical. An impulsive force J is applied at a point on the axis of symmetry at a distance d from P and in a direction perpendicular to the vertical plane through the axis of symmetry. Calculate the change in the kinetic energy immediately after the impulse and find the impulsive reaction at P.

From Chapter 8 we know that the kinetic energy of a top is given by

$$T = \tfrac{1}{2}A\dot{\theta}^2 + \tfrac{1}{2}A\dot{\phi}^2 \sin^2 \theta + \tfrac{1}{2}C(\dot{\psi} + \dot{\phi}\cos\theta)^2$$

where θ, ϕ, ψ are Eulerian angles. In steady precession $\dot{\theta} = 0$, $\theta = \alpha$,

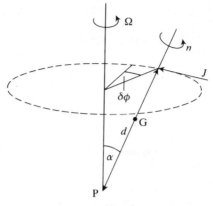

Fig. 9.4

$\dot{\phi} = \Omega$, and $\dot{\psi} + \dot{\phi} \cos \theta = n$. In this example we have a well-known set of generalized coordinates and it turns out to be easiest to calculate the generalized impulsive forces using these as in eqn (9.21). We look at virtual displacements $\delta\theta$, $\delta\phi$, $\delta\psi$. The displacement $\delta\psi$ is just a change in the orientation about the axis of symmetry and so has no effect. The displacement $\delta\theta$ changes the angle of inclination to the vertical and so the movement of each point on the axis is perpendicular to J as it occurs in the vertical plane through the axis. Only the displacement through $\delta\phi$ produces a contribution. The distance moved by the point of application of J is $d \sin \alpha(\delta\phi)$ perpendicular to the vertical plane. Hence (9.21) gives

$$J d \sin \alpha(\delta\phi) = J_\theta \delta\theta + J_\phi \delta\phi + J_\psi \delta\psi$$

so that $J_\theta = J_\psi = 0$ and $J_\phi = J d \sin \alpha$. Note that the impulsive reaction at P does no work in a virtual displacement because P is stationary.

Using eqns (9.20)

$$\left[\frac{\partial T}{\partial \dot{\theta}} \right]_1^2 = J_\theta \Rightarrow A\dot{\theta} - 0 = 0$$

for $\dot{\psi}$

$$C(\dot{\psi} + \dot{\phi} \cos \alpha) - Cn = 0$$

and for $\dot{\phi}$

$$A\dot{\phi} \sin^2 \alpha + C \cos \alpha(\dot{\psi} + \dot{\phi} \cos \alpha) - A\Omega \sin^2 \alpha - Cn \cos \alpha = J d \sin \alpha$$

since $\theta = \alpha$. The equation for $\dot{\psi}$ shows that the spin about the axis is unchanged. Substituting in the equation for $\dot{\phi}$ gives

$$A\dot{\phi}\sin^2\alpha = Jd\sin\alpha + A\Omega\sin^2\alpha$$

$$\dot{\phi} = \frac{Jd}{A\sin\alpha} + \Omega.$$

The change in kinetic energy is given by

$$\Delta T = \frac{1}{2}A\left(\frac{Jd}{A\sin\alpha} + \Omega\right)^2 \sin^2\alpha + \tfrac{1}{2}Cn^2 - (\tfrac{1}{2}A\Omega^2\sin^2\alpha + \tfrac{1}{2}Cn^2)$$

$$= Jd\Omega\sin\alpha + \frac{J^2d^2}{2A}.$$

Note that, as we might have expected, if J has the same sense as Ω then the kinetic energy increases, and if J is in the opposite direction to Ω there is an increase in the kinetic energy provided

$$|J|d > 2A\Omega\sin\alpha.$$

Otherwise the kinetic energy decreases.

To calculate the impulsive reaction at P we must use eqn (9.2) involving the change in linear momentum (Section 9.1). Before the impulse the velocity of the centre of mass, G, is $\Omega h\sin\alpha\mathbf{j}$ where $\mathbf{j}$ is a unit vector perpendicular to the vertical plane through the axis of symmetry and h is the distance of G from P. Immediately after the impulse the velocity is

$$\left(\frac{Jd}{A\sin\alpha} + \Omega\right)h\sin\alpha\mathbf{j}.$$

Using the equation $M(\mathbf{v}_2 - \mathbf{v}_1) = \Sigma\,\mathbf{J}$ gives

$$M\frac{Jd}{A\sin\alpha}h\sin\alpha\mathbf{j} = J\mathbf{j} + \mathbf{R}$$

where $\mathbf{R}$ is the impulsive reaction at P. This equation establishes that there will usually be an impulsive reaction at P given by

$$\mathbf{R} = J\left(\frac{Mdh}{A} - 1\right)\mathbf{j}.$$

Sections 9.1 and 9.2 show that the combination of Newtonian and Lagrangian methods can provide the means to solve problems in impulsive motion of mechanical systems. We have looked at examples where analytic solutions or partial analytic solutions can be found. The models outlined can be used to solve very complicated problems numerically using a computer, once the model for the mechanical system involved is properly formulated.

Exercises:
Chapter 9

1. Two small spheres mass m_1, m_2 are moving along the same straight line. They collide when their velocities are u_1, u_2. Immediately after the collision the velocities are v_1, v_2. Newton's experimental law states that

$$v_2 - v_1 = -e(u_2 - u_1)$$

where the coefficient of restitution e satisfies $0 < e \leq 1$. Find v_1 and v_2 and show that the kinetic energy decreases in general, remaining the same only if $e = 1$.

2. A rocket sits on a launch pad. The mass of the rocket before the motors are switched on is M and the rate at which fuel is burnt is a constant $k > 0$. The exhaust gases are emitted at constant speed u relative to the rocket. For one-dimensional vertical motion only, write down the variable mass equation for the rocket on the launch pad as the motors are switched on, and show that it takes off immediately if $ku > Mg$. Assuming that this inequality holds, find the velocity of the rocket at a time t when the rocket is still close to the surface of the Earth.

   If the rocket's motors are still firing at some distance from the Earth write down the equation of motion using an inertial frame at the centre of the Earth. Assume that the velocity of the exhaust gases relative to the rocket directly opposes the velocity of the rocket itself in the inertial frame. Prove that the triple scalar product $(\ddot{\mathbf{r}}, \dot{\mathbf{r}}, \mathbf{r}) = 0$, where $\mathbf{r}$ is the position vector of the rocket from the centre of the Earth. What does this tell you about the path of the rocket under these assumptions?

3. A snooker ball, mass m, radius a, is at rest on a table. A player strikes it with a cue in such a way that a horizontal impulsive force $\mathbf{J}$ acts on the ball, applied at the point with position vector $\mathbf{r}$ from its centre. The force $\mathbf{J}$ and the vector $\mathbf{r}$ lie in the same vertical plane through the centre of the ball. Find the range of vector values which can be taken by the velocity of the point of contact, as $\mathbf{r}$ varies from the top to the bottom of the ball. If $\mathbf{r}$ is no longer necessarily in the vertical plane through G, find also the maximum and minmum values of the kinetic energy acquired and the corresponding positions $\mathbf{r}$.

4. A signboard in the form of a uniform rectangle, with edges of length $2a$, $2b$ and mass m, is spinning with angular speed ω about a line through the centre of mass parallel to the edges of length $2b$ when a vertex is suddenly fixed. Calculate the new angular velocity and the loss of kinetic energy.

5. A top, which has moments of inertia A, A, C about principal axes at its vertex P, is freely pivoted at P and spinning with angular speed ne where $\mathbf{e}$ is a unit vector along the axis of symmetry. An impulse $\mathbf{J}$, perpendicular to $\mathbf{e}$, is applied at a point Q whose position vector from P is $\mathbf{d}$, making an acute angle α with $\mathbf{e}$. The impulse $\mathbf{J}$ is also perpendicular to $\mathbf{d}$. Calculate the direction of the new angular velocity and the change in kinetic energy.

6. Three uniform identical rods AB, BC, CD are lying at rest on a table along three sides of a square. They are freely jointed at B and C. An impulse J, parallel to the table, is applied at A in the direction directly away from D. Show that the distance DA tends to lengthen initially and find the kinetic energy.

Suppose now that the rod CD slides through a ring at its midpoint which is attached to the table but which can freely rotate. What constraint does this place on the velocities immediately after the impulse? Find the kinetic energy in this case.

7. A mechanical system has kinetic energy which is expressed as a homogeneous quadratic in terms of the generalized velocities $\dot{q}_i$. Show that if it is acted on by impulsive forces $\hat{Q}_i$ then the change in kinetic energy is given by

$$\Delta T = \tfrac{1}{2} \sum \hat{Q}_i(u_i + v_i)$$

where the generalized velocity $\dot{q}_i$ takes the values u_i, v_i before and after the impulse respectively.

Find the change of kinetic energy in example 4 in the text, checking that the formula works in this case.

APPENDICES

1. Conics

The aim of this section is to enable those of you who have not studied conics to understand the differing trajectories of astronomical bodies (see Chapter 2). We can visualize conics using a cone (a wizard's hat!) intersected by a plane. Depending on the angle of the plane to the axis of symmetry of the cone, the curve of intersection is either an ellipse (a closed curve) or a hyperbola or parabola (both infinite curves). The parabola occurs in the special case when the plane is parallel to one of the straight lines in the cone which go through the vertex (a generator/edge looking at the cone from a side view). The equations of these curves are shown below in their standard form.

ELLIPSE

Equation:

$$\frac{x^2}{a^2} + \frac{y^2}{b^2} = 1.$$

For $a > b$, the x axis is the major axis, the y axis is the minor axis.

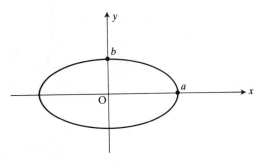

Fig. A1

HYPERBOLA

Equation:

$$\frac{x^2}{a^2} - \frac{y^2}{b^2} = 1.$$

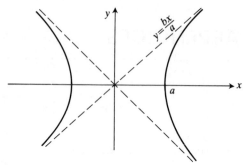

Fig. A2

PARABOLA

Equation:

$$y^2 = 4ax.$$

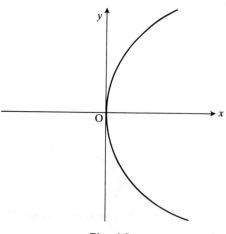

Fig. A3

The ellipse, hyperbola, and parabola are the only distinct shapes which result from the intersection of a plane and a cone. The circle is a special case of the ellipse, given above by $a = b$. In the diagrams above the centre of each conic is the origin and the curves are symmetrical under a reflection in the x axis. By rotating the axes Oxy or by translating the origin O we can get very different equations but we do not change the shapes of the curves.

We are interested in the polar form of the conic equation. The path of a body moving under the influence of the gravitational inverse square law force is naturally found in this form. The first step is to find a different way of deriving the equation of a conic. We take a set of axes Oxy and look at the point $(ae, 0)$, the focus, and the line $x = a/e$, the directrix. The non-negative number e is called the eccentricity. We will assume first of all that $0 < e < 1$. Then we draw on the graph every point whose distance from the focus is e times its distance from the directrix.

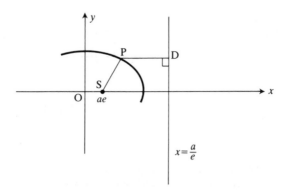

Fig. A4

From the diagram, $PS = e\,PD$. If P is the point (x, y) this gives

$$(x - ae)^2 + y^2 = e^2 \left(\frac{a}{e} - x \right)^2.$$

Multiplying out

$$x^2(1 - e^2) + y^2 = a^2(1 - e^2) = b^2$$

defining b, and hence

$$\frac{x^2}{a^2} + \frac{y^2}{b^2} = 1.$$

Try the same procedure assuming $e > 1$ and the equation of a hyperbola emerges, the only difference being that now $b^2 = a^2(e^2 - 1)$. In both cases it is clear that the same equation would have resulted had we used a point $(-ae, 0)$ and a line $x = -a/e$. The ellipse and the hyperbola each have two foci $(\pm ae, 0)$ and two directrices $x = \pm a/e$.

To find a parabola in this way we must put $e = 1$. This places the focus $(a, 0)$ on the directrix $x = a$ and the method breaks down. However, the equation of the parabola emerges if we use the focus $(a, 0)$ and the opposing directrix $x = -a$ instead.

THE POLAR EQUATION

Firstly we shift the origin of the coordinates to a focus, $(ae, 0)$. The directrix then becomes $x = (a/e) - ae$. Again we assume that $0 < e < 1$.

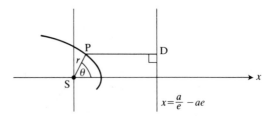

Fig. A5

We intend to use plane polars so that r is the distance of P from the origin S and θ is the angle between SP and the x axis

$$x = r \cos \theta \qquad y = r \sin \theta.$$

As before $PS = e PD$. Hence

$$r = e \left(\frac{a}{e} - ae - r \cos \theta \right).$$

Setting $p = a(1 - e^2)$ we have

(∗)
$$\frac{p}{r} = (1 + e \cos \theta).$$

It is easy to check that using the other focus and directrix simply changes $+$ to $-$ on the right-hand side. This is the polar form of the conic equation. The same equation arises for the hyperbola, taking $p = a(e^2 - 1)$, and for the parabola, taking $p = 2a$ and $e = 1$. The main point to watch out for is that the origin is not at the centre of the conic, but at a focus.

We now have the polar equation of a conic (∗), from which calculating the eccentricity e easily classifies the conic:

(1) an ellipse, $0 \le e < 1$, a circle, $e = 0$;

(2) a parabola, $e = 1$;

(3) a hyperbola, $e > 1$.

2. Cylindrical polar coordinates

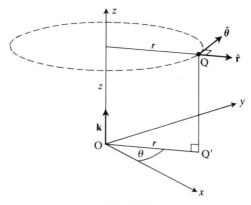

Fig. A6

Suppose that in Cartesian coordinates the point (x, y, z) represents Q in the diagram. Cylindrical polar coordinates are plane polar coordinates with the addition of the z coordinate, (r, θ, z) replacing (x, y, z) where

$$x = r\cos\theta \qquad y = r\sin\theta \qquad z = z.$$

The coordinate r is the distance of Q from the Oz axis. We use $\hat{\mathbf{r}}$ as a unit vector in the direction of increasing r and $\hat{\boldsymbol{\theta}}$ as a unit vector in the direction of increasing θ. This means that $\hat{\boldsymbol{\theta}}$ is the tangent to a circle around the Oz axis as shown. Then the vector $\underline{OQ} = \mathbf{q}$ is given by

$$\mathbf{q} = r\hat{\mathbf{r}} + z\mathbf{k}$$

where $\mathbf{k}$ is a unit vector along the z axis as usual. In cylindrical polar coordinates the velocity is

$$\dot{\mathbf{q}} = \dot{r}\hat{\mathbf{r}} + r\dot{\theta}\hat{\boldsymbol{\theta}} + \dot{z}\mathbf{k}$$

and the acceleration is

$$\ddot{\mathbf{q}} = (\ddot{r} - r\dot{\theta}^2)\hat{\mathbf{r}} + \frac{1}{r}\frac{d}{dt}(r^2\dot{\theta})\hat{\boldsymbol{\theta}} + \ddot{z}\mathbf{k}.$$

3. Spherical polar coordinates

In the system of spherical polar coordinates each point Q, (x, y, z), is represented by its distance from the origin from the axes $Oxyz$ and by two further angles θ, ϕ.

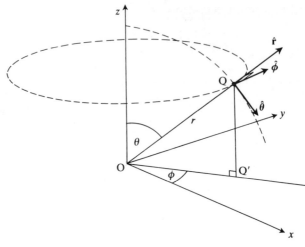

Fig. A7

The coordinate r is the distance OQ, the coordinate θ is the angle between OQ and the z axis, and the coordinate ϕ is the angle between the plane containing Oz and OQ and the plane Oxz. In terms of spherical polar coordinates, Cartesian coordinates can be expressed neatly as

$$x = r\sin\theta\cos\phi \qquad y = r\sin\theta\sin\phi \qquad z = r\cos\theta.$$

The unit vector $\hat{\mathbf{r}}$ is along OQ extended. The unit vector $\hat{\boldsymbol{\theta}}$ is in the direction of increasing θ and similarly $\hat{\boldsymbol{\phi}}$ is in the direction of increasing ϕ. If you imagine the point Q on a globe with the Oz axis along the Earth's polar axis pointing north, then $\hat{\boldsymbol{\theta}}$ is tangential to the circle of longitude at Q and $\hat{\boldsymbol{\phi}}$ is tangential to the circle of latitude at Q. The position vector is given by OQ $= \mathbf{r} = r\hat{\mathbf{r}}$, and the velocity by

$$\dot{\mathbf{r}} = \dot{r}\hat{\mathbf{r}} + r\dot{\theta}\hat{\boldsymbol{\theta}} + r\sin\theta\dot{\phi}\hat{\boldsymbol{\phi}}.$$

The unit vectors are expressed as follows:

$$\hat{\mathbf{r}} = \sin\theta\cos\phi\,\mathbf{i} + \sin\theta\sin\phi\,\mathbf{j} + \cos\theta\,\mathbf{k}$$

$$\hat{\boldsymbol{\theta}} = \cos\theta\cos\phi\,\mathbf{i} + \cos\theta\sin\phi\,\mathbf{j} - \sin\theta\,\mathbf{k}$$

$$\hat{\boldsymbol{\phi}} = -\sin\phi\,\mathbf{i} + \cos\phi\,\mathbf{j}.$$

SOLUTIONS TO ODD-NUMBERED EXERCISES

Chapter 1

1. Maximum speed $\frac{1}{4}a\sqrt{(k/m)}$; period $2\pi\sqrt{(m/k)}$.
3. The bob moves in a circle so that the only component of acceleration is towards the centre.
5. (a) 11.8 m; (b) maximum range is $V^2/g(1 - \sin \beta)$ when $\alpha = \pi/4 + \beta/2$.
7. If $\mu = 0$ then $\ddot{\mathbf{r}} = g \sin \alpha\, \mathbf{i}$. Choosing $\mathbf{r} = \mathbf{0}$ at $t = 0$, then $x = \frac{1}{2}gt^2 \sin \alpha + ut$ down the plane and $y = vt$ horizontally in the plane, u, v being constant. Hence the path is parabolic unless $v = 0$.

Chapter 2

1. At $t = 0$ choose $\theta = 0$. Then $\dot{r} = 0$, $r\dot{\theta} = V$, $r = R$, and $h = r^2\dot{\theta} = RV$. Since the speed is increased to V, $RV^2 > GM$. For continuing orbit $RV^2 < 2GM$ and for escape $RV^2 \geq 2GM$. The orbit must be elliptic. The escape path is part of a hyperbola unless equality holds when it is part of a parabola. Minimum escape velocity is 8.9×10^3 m s^{-1}.
3. The angle between the maximum and minimum distance is $\frac{1}{2}\pi$.
7. The vectors $\mathbf{r}, \dot{\mathbf{r}}, \ddot{\mathbf{r}}$ are coplanar $\Rightarrow$ planar motion.

Chapter 3

3. The distance of closest approach is given by

$$\frac{\gamma + \sqrt{(\gamma^2 + d^2 V^4)}}{V^2}.$$

Note this expression tends to d as $\gamma \to 0$.
5. The energy equation leads to

$$\left(1 + \frac{a^4}{z^4}\right)\dot{z}^2 = -\omega^2(z - a)\left(z - \frac{2g}{\omega^2} + a\right).$$

Motion must lie between the roots and is therefore circular if they coincide.

7. The minimum value of the potential energy is $-e_0$ when $r = (^6\sqrt{2})a$. Kinetic energy is positive and so minimum possible energy is $-e_0$. For motion with $h = 0$ then $\dot{\theta} = 0$ and a minimum or maximum distance of the particle from O occurs when $\dot{r} = 0 \Rightarrow$

$$\left(\frac{a}{r}\right)^6 = \frac{1}{2} \pm \sqrt{\left(\frac{1}{4} + \frac{E}{4e_0}\right)}.$$

Chapter 4

1. Angular velocity $\dot{\theta}\mathbf{k} + \Omega\mathbf{e}$; proposition 4.2 states angular velocities are additive.
3. Period $2\pi\sqrt{(a/(g - a\omega^2))}$.
5. Solution to first paragraph is identical to eqn (4.16).
7. Solutions to the equations in the rotating frame are:

$$x = \frac{2c}{\omega} + d\cos(\omega t + \varepsilon) \quad y = b - 3ct - 2d\sin(\omega t + \varepsilon)$$

$$z = A\cos(\omega t + \delta)$$

where b, c, d, A, ε, and δ are constants. When $x = y = 0$, the solution for a small change in z alone is harmonic and hence stable.

Chapter 5

1. If $r = O'P_2$ and $u = r^{-1}$ then the differential equation satisfied by u is

$$\frac{d^2u}{d\theta^2} + u = \frac{1}{a}$$

and measuring $\theta = 0$ from the initial direction of P_1P_2, at $\theta = 0$, $u = 1/a$, $du/d\theta = \sqrt{(3)}/a$. Solving

$$u = \frac{1}{a} + \frac{\sqrt{3}}{a}\sin\theta.$$

A similar equation can be found for P_1.

Chapter 6

3. The equivalent force and couple are $\mathbf{F}$ at the point with position vector $\mathbf{a}$ and a couple $\mathbf{\Gamma} - \mathbf{a} \wedge \mathbf{F}$.

5. (a) $\frac{2}{5}ma^2 I$; (b) taking axes with two perpendicular to the axis of symmetry and one along the axis then $\mathfrak{I}_G = \text{diag}\{A, A, C\}$ with $A = M(\frac{1}{4}a^2 + \frac{1}{12}h^2)$ and $C = \frac{1}{2}Ma^2$; (c) taking axes Ox, Oy, parallel to the sides of length $2a$, $2b$ respectively and Oz perpendicular to the lamina $\mathfrak{I}_G = \text{diag}\{A, B, C\}$ with $A = \frac{1}{3}Mb^2$, $B = \frac{1}{3}Ma^2$, and $C = \frac{1}{3}M(a^2 + b^2)$.

7. The principal moments of inertia are $\frac{2}{3}Ma^2$ and $\frac{11}{3}Ma^2$ twice. The principal axes are the axis joining the centre of mass to the vertex and any axis perpendicular to it.

9. If Qx is the tangent to the rim of the base, Qy is along the radius to Q in the base, and Qz is perpendicular to the base then the points $(\pm x, y, z)$ are either both inside or both outside the cone. Hence $G = 0 = H$ and Qx is a principal axis. If in addition the cone has dynamical symmetry then a second axis is from Q to the centre of mass and the third must be perpendicular to these two.

Chapter 7

1.

$$\omega = \frac{1}{\lambda}\mathbf{G}_0 + \left(\omega_0 - \frac{1}{\lambda}\mathbf{G}_0\right)\exp[(-\lambda/A)t] \to \frac{1}{\lambda}\mathbf{G}_0 \text{ as } t \to \infty.$$

5. Look at the vector equation

$$A\mathbf{e} \wedge \ddot{\mathbf{e}} + Cn\dot{\mathbf{e}} + C\dot{n}\mathbf{e} = -Mghe \wedge \mathbf{k}$$

and take its scalar product with $\mathbf{e}$, $\mathbf{k}$, $\mathbf{e} \wedge \dot{\mathbf{e}}$ integrating each expression once to get the required constants of the motion.

7. There are two real solutions of the equation involving Ω, for each pair of values α, n. The path traced out by the point of contact is circular.

9. The acceleration of the centre of mass is $\mathbf{0}$ and so it must move in a straight line.

Chapter 8

1.

$$E = \frac{1}{2}m(\dot{r}^2 + r^2\dot{\theta}^2) - \frac{GMm}{r} \quad \text{and} \quad \frac{\text{d}E}{\text{d}t} = 0.$$

3.

$$L = \frac{m}{6}(3\dot{x}^2 + 3a\dot{x}\dot{\theta}\cos\theta + a^2\dot{\theta}^2) + \frac{1}{2}mga\cos\theta$$

where x is the distance of the ring along the wire measured from a fixed origin and a is the length of the rod, and

$$a\ddot{\theta}(\tfrac{1}{3} - \tfrac{1}{4}\cos^2\theta) + \tfrac{1}{4}a\dot{\theta}^2 \sin\theta\cos\theta + \tfrac{1}{2}g\sin\theta = 0.$$

5. If $2\sqrt{(2)}a\Omega^2 = g$ the bob moves in a circle, i.e. we have $\theta = \tfrac{1}{4}\pi \; \forall t$.

7. The normal coordinates are scalar multiples of the extension beyond the equilibrium length and the angle between the string and the downward vertical.

Chapter 9

1. $(m_1 + m_2)v_1 = (m_1 - em_2)u_1 + (m_2 + em_2)u_2$, similarly v_2. The loss in energy is given by

$$\frac{1}{2}\frac{m_1 m_2}{(m_1 + m_2)}(1 - e^2)(u_1 - u_2)^2.$$

3. The velocity of the point of contact is $(\lambda/m)\mathbf{J}$ where $-\tfrac{3}{2} < \lambda < \tfrac{7}{2}$. The maximum possible energy is $\tfrac{7}{4}\mathbf{J}^2/m$ which is almost achievable if the blow is struck at a point as close as possible to the circle on the surface of the ball which is in the plane perpendicular to $\mathbf{J}$. The minimum energy is $\mathbf{J}^2/2m$. This is the result if the blow is struck at the point Q of the ball such that QG and $\mathbf{J}$ are parallel.

5. Angular velocity:

$$\frac{Jd}{A}\cos\alpha\,\mathbf{i} + \left(n + \frac{Jd}{C}\sin\alpha\right)\mathbf{e}$$

where $\mathbf{i}$ is a unit vector perpendicular to $\mathbf{e}$ and in the same plane as $\mathbf{e}, \mathbf{d}$. Kinetic energy:

$$\frac{1}{2}\frac{J^2 d^2}{A}\cos^2\alpha + \frac{1}{2}C\left(n + \frac{Jd}{C}\sin\alpha\right)^2.$$

7. Since $J_\theta = 0 = J_\psi$ we must have

$$\Delta T = \tfrac{1}{2}J_\phi(\dot{\phi}_1 + \dot{\phi}_2)$$

where the subscript 1 and 2 refer to immediately before and after the impulse. Hence the change in kinetic energy is

$$\Delta T = J\Omega d\sin\alpha + \frac{J^2 d^2}{2A}.$$

INDEX